Shale Gas

Shale Gas
Exploration and Environmental and Economic Impacts

Edited by

Anurodh Mohan Dayal
Devleena Mani
CSIR-NGRI, Hyderabad, India

ELSEVIER   elsevier.com

Library of Congress Cataloging-in-Publication Data
A catalog record for this book is available from the Library of Congress

British Library Cataloguing-in-Publication Data
A catalogue record for this book is available from the British Library

ISBN: 978-0-12-809573-7

For information on all Elsevier publications
visit our website at https://www.elsevier.com/books-and-journals

Working together
to grow libraries in
developing countries

www.elsevier.com • www.bookaid.org

Publisher: Candice Janco
Acquisition Editor: Amy Shapiro
Editorial Project Manager: Tasha Frank
Production Project Manager: Anitha Sivaraj
Designer: Vicky Pearson Esser

Typeset by TNQ Books and Journals

Dedicated to

all our family members

Contents

List of Contributors

A.M. Dayal   CSIR-NGRI, Hyderabad, India
M.S. Kalpana   CSIR-NGRI, Hyderabad, India
D. Mani   CSIR-NGRI, Hyderabad, India
D.J. Patil   CSIR-NGRI, Hyderabad, India
U. Vadapalli   CSIR-NGRI, Hyderabad, India
A.K. Varma   Indian School of Mines, Dhanbad, Jharkhand, India
N. Vedanti   CSIR-NGRI, Hyderabad, India

Foreword

Shale gas and shale oil have been considered a new generation, unconventional resource of natural gas and oil. These nonconventional sources of natural gas and oil are capable of increasing production to a much higher level. Shale gas is seen as a major game changer for the global petroleum industry. Generation of crude oil and natural gas in sedimentary basins takes place in poorly permeable shales, argillaceous calcareous rocks, or in the intrashelf micritic limestones, which have been named as source rocks. They are the targets for exploitation of unconventional shale gas and/or oil in contrast to the conventional gas and oil that can be so produced from oil and gas wells without any drastic artificial increase of porosity and permeability of the host rock.

The shale gas is a very good source of nonconventional hydrocarbon resources. Major players in shale gas plays are the United States, Canada, China, and Arizona. Though hydraulic fracturing was first carried out in 1949, for the exploitation of shale gas the development of hydraulic fracturing and horizontal drilling has been developed in the last 15 years in the United States. A new era to meet the energy demand through production of shale oil and shale gas had begun, with preference for the latter because of many upcoming players. Brazil and Russia are going to be shale gas producers in the near future.

Today, the United States, which is practically the sole owner of the massive hydrofracturing technology, produces more unconventional gas per day than conventional, meeting around 46% of the domestic demand and making the United States fully self-reliant for domestic gas availability and an exporter of Liquefied Natural Gas (LNG). China and Canada are attempting to be like the United States, while operations by the newer players are yet to evolve into large-scale operations.

The unconventional oil and gas resources offer opportunities to many countries to minimize their dependence on the Middle East to meet their ever-growing energy demand in consonance with the needs of economic development and to even become net oil or gas exporters. The list of countries that will engage in shale oil and gas exploitation will surely grow.

Petroleum geochemistry provides proven methodologies to identify the best potential shale sequences. The combined use of modern petrophysical techniques and surface seismic methods helps to delineate aerial extents of friable, organic matter–rich shale units possessing the best potential for substantial oil and gas production on fracturing. Persistent research is for developing better fractured shale reservoir models to better plan and monitor shale oil and gas production from fractured reservoirs.

With the constant improvement in geoscientific methods to map shale units with the best shale oil and gas potential and the fracturing technologies that include a set of waterless cost-effective fluid fracturing technologies along with advent of more players providing massive fracturing services, the shale gas industry is bound to grow. Most of the environmental hazards that have been observed to be associated with or propagated about fracturing will be fully overcome to the best satisfaction of various regulatory authorities for the shale oil and gas industry. The current depressed crude oil and natural gas prices are bound to provide impetus for research to make the shale oil and gas industry highly cost competitive. With the assured growth of the shale oil and gas industry despite vagaries of fluctuations of prices of crude oil and natural gas, the industry will demand a constant supply of petroleum scientists and engineers who are well equipped with knowledge of shale oil and gas exploration and exploitation. The shale gas industry shall continue to hold a priority, as natural gas is a more environmentally friendly energy source compared to any other organic fossil fuel, having the lowest carbon fingerprint, with a promise to minimize the rate of global warming and rate of climate change.

The present book of Dr. A.M. Dayal and his coworker Dr. Devleena Mani is very timely, as more and more countries are vying to be listed as shale gas–producing countries. The book fills the need from industry as well as from academia to eruditely and comprehensively cover the subject of shale gas resource appraisal and shale gas exploitation to the benefit of new entrants in the shale gas industry.

Dr. A.M. Dayal is an emeritus scientist of the Council of Scientific and Industrial Research (CSIR) and his team of scientists at the Center of Excellence in Geochemical Prospecting at the CSIR-National Geophysical Research Institute (NGRI), Hyderabad, supported by Directorate General of Hydrocarbons and Oil Industry Development Board, Ministry of Petroleum and Natural Gas, Government of India, are engaged with the project on characterizing shale gas potentials of the Cambay Basin of India from 2012. His coworkers are currently involved with the project work on other Indian sedimentary basins. The book authored by Dr. A.M. Dayal and Dr. Devleena Mani is assured to be a storehouse of very useful and practical knowledge on shale gas potential appraisal and shale gas exploitation methodologies.

I do foresee that this book will be an inalienable addition to the literature on shale gas exploration and exploitation and will be kept in the libraries of oil and gas companies and libraries of various universities, the world over, and it will benefit a large number of students, research geologists, research geochemists, research geophysicists, and technologists who aim to be soldiers for global energy security through hydrocarbons.

Dr. Kuldeep Chandra
Former Executive Director R&D
KDMIPE, ONGC
Dehradun
India

Preface

Increasing requirements and diminishing resources of fossil fuel in the last two decades have triggered the need for alternate sources of energy, which shall also have a minimum impact on the environment. The presence of gas in shale formations was known much earlier, but it was the development of horizontal drilling and hydraulic fracturing, when advanced sufficiently enough, that resulted in the exploration and exploitation of shale gas. The United States took the lead, and the combined efforts of industry and academia led to the success of shale gas production on a commercial scale in a very large amount. This was also responsible for the downtrend of gas prices from $8–10 to $2.5 per million Btu in last 5 years. An important advantage of using shale gas as a fuel is that there is no particulate matter nor carbon dioxide emission. There is methane emission, however, which can be controlled.

The book *Shale Gas: Exploration, Environmental and Economic Impacts*, edited by Drs. Anurodh Mohan Dayal and Devleena Mani, presents a precise yet elaborate view of various aspects associated with shale gas. It provides the necessary details on the geology, mineralogy, petrochemistry of shale rocks, the gas generation and storage, and the technology required to harness it, as well as its environmental and economic impacts. The book is useful to graduate and postgraduate students, academicians, industrialists, economists, and also those not belonging to these backgrounds owing to the simplicity with which the book has been sketched.

The topics on geology of shale, deposition and diagenesis, source and reservoir properties, basin structure and tectonics, exploration technique, hydraulic fracturing, impact to the environment and economics, and global demand in the future of shale gas have been covered in 10 chapters. The fundamentals of shale rocks, its physical and chemical composition, and its role as a source and reservoir rock for natural gas have been discussed in Chapter 1. Geological and geochemical conditions for the deposition of organic-rich sediments are important for the generation of gas and augment shale's potential to release the gas on application of suitable fracking methods. Chapter 2 describes the depositional and diagenetic conditions of shale rocks. Sedimentary organic matter is the source of gaseous hydrocarbons in shales. The organic content and composition with suitable thermal maturity governs the generation potential of shale gas. The origin and evolution of organic matter to hydrocarbons under the influence of temperature and pressure in discussed in Chapter 3. It also includes an account of a diverse array of geochemical tools used to characterize the organic matter in sediments such as the bulk and molecular-level organic and isotopic signatures of organic compounds that are vital to gas generation. Whereas the carbon isotopic study of organic matter provides information on the origin of gas, the pyrolysis of the shales describes the type of kerogen associated with the shale formation. The organic facies is composed of microscopic organic materials or macerals, derived from terrestrial, lacustrine, and marine plant remains and are a function of the parent source material. These macerals, upon exposure to higher temperatures in the subsurface, predominantly produce gaseous hydrocarbons. Thermally mature organic matter generates oil, whereas the postmature organic matter is in the wet and dry gas zones.

The basin structure, tectonics, and stratigraphy are vital to the deposition and preservation of organic matter and also control its maturation to natural gas. Based on the characteristic geologic settings of the basin and the shale formation of interest, the fracking and fracturing techniques are relegated. Chapter 4 covers the fundamentals of sedimentary basins and role of tectonics and stratigraphy in defining a commercial shale gas play.

Successful exploration of shale gas involves the employment of geological, geophysical, and geochemical concepts, which have been dealt in Chapter 5. Shallow seismic study of the basin is necessary to know the extent and thickness of shale formations. Geochemistry of shale is an important tool for the exploration. The Total Organic carbon (TOC) content, thermal maturity, and kerogen properties provide useful information on its gas generation potential. Petrophysical characterization of shale rocks involves results from gamma ray study, resistivity, and porosity along with neutron capture spectroscopy data. Shales commonly contain higher levels of naturally occurring radioactive materials such as thorium, potassium, and uranium. Gamma ray logging provides one of the first indications of the presence of organic-rich shale. Porosity measurements also have specific characteristics for organic-rich shales, which exhibit higher density porosity and lower neutron porosity, owing to the presence of gas in the rock. For the characterization of unconventional reservoirs, evaluation of shale formation depends on understanding the mineralogy of rocks. Acoustic measurements, especially those that provide mechanical properties for anisotropic shale media, are also significant for understanding the long-term productivity of shale gas wells. Sonic porosity is another acoustic measurement that is beneficial in shale analysis. Higher sonic porosity than the neutron porosity indicates the presence of gas in the pore spaces.

Different clay minerals or phyllosilicates having leaf or plate-like structure are responsible for the fissility of shale rocks. Illite, mixed layer illite/semectite, semectite, kaolinite, and chlorite are the dominant clay minerals in shales. Shale lithology and mineral composition are the main intrinsic factors controlling fracture development in shale.

Chapter 6 discusses the hydraulic fracturing process for shale gas exploration. It involves the injection of fluids into a subsurface geologic formation containing oil and/or gas at a high pressure sufficient to induce fractures through which oil or natural gas can flow to a producing wellbore. A large quantity of water is required for hydraulic fracturing, the management and disposal of which is one of the biggest challenges. The expansion of the fractures depends on the reservoir and rock properties, boundaries above and below the target zone, the rate at which the fluid is pumped, the total volume of fluid pumped, and the viscosity of the fluid. In addition to water and proppant, other additives are essential to successful fracture stimulation.

Case studies for the production and potential countries have been discussed in Chapter 7. According to Energy Information Administration (EIA), the production of shale gas in the United States was 11.34 Tcf in 2013, which is 47% of the total natural gas production. Argentina, Canada, and China are the other countries producing shale gas. In developing countries, India (625 Tcf), Pakistan (1000 Tcf), Indonesia (574 Tcf), Australia (437 Tcf), Africa (485 Tcf), and Russia (1920 Tcf) have good reserves of shale gas.

Shale gas exploitation and the environment are interrelated issues. The emission of air pollutants from shale gas are in large-scale compared to conventional gas exploration and include emissions from diesel generators, leakage of methane from well pads, volatile organic compounds, particulate matter, and high-level noise production. Volatile organic carbon and nitrous oxide released during the production together form smog in the atmosphere. Induced seismicity is reported from operations related to shallow seismic and hydraulic fracturing. Careful handling of the chemicals as additives is another challenge, particularly the carcinogenic elements in the flow water. Surface blowouts are dangerous for the workers and also contamination of drilling fluid with surface water. All these have a direct impact on living beings, and their preventive measures and remedial technologies are covered.

Presently, the United States, China, Canada, and Argentina are exploiting shale gas. The United States is now self-sufficient for gas and oil and exports to India, the European Union, and Latin America. Shale gas has changed the global economics of the oil and gas industry. Chapter 9 discusses role of shale gas as a game changer for the global economy. Increasing gaps in demand and supply of conventional hydrocarbon have resulted in increasing rates of hydrocarbons. The economic growth of the developing countries depends on the consumption per capita of hydrocarbon. Shale gas production paved the way to the downfall of global prices of hydrocarbon from \$110 to \$40 per barrel, thus reducing the economic crisis for developing countries like India, China, South Asia, and Africa to a certain extent and also resulting in growth of infrastructure and the economy.

It is expected that with many advantages, shale gas will replace conventional hydrocarbon and coal as a main source of energy in the next century. Shale gas has been substituted for coal in a number of powerhouses in the United States due to low carbon dioxide emission, so net carbon emission is also reduced. South America and China are also planning to use shale gas as a main source of energy. With increasing prices of oil in global market, the investment in shale gas will be quite encouraging.

Dr. Anurodh Mohan Dayal
Dr. Devleena Mani

Acknowledgements

Dr. Anurodh Mohan Dayal acknowledges the Director General, Council of Scientific and Industrial Research (CSIR), New Delhi, for the award of Emeritus Scientist position. Dr. Devleena Mani is thankful to CSIR for funding the Senior Research Associateship. The Director, CSIR-National Geophysical Research Institute, Hyderabad, is acknowledged for permitting the publication of this work. The authors gratefully acknowledge the scientific, technical, and research fellows of Petroleum Geochemistry lab, CSIR-NGRI, for their support while writing this book. We also thank other staff of our institute who supported us while writing this book.

Special thanks are due to the Secretary, Oil Industry Development Board (OIDB), New Delhi, for providing financial aid in setting up the State of Art Petroleum Geochemistry lab laboratory at CSIR-NGRI, Hyderabad. Officials of Central Mine Planning & Design Institute (CMPDI), Ranchi and Singareni Collieries Company Ltd. (SCCL), Kothagudem, are thanked for extending their support during field visits. Dr. Nimisha Vedanti acknowledges the Director, Exploration, Oil Natural Gas Company (ONGC) for providing the data for her work.

Dr. Anurodh Mohan Dayal
Dr. Devleena Mani

CHAPTER 1

Shale

A.M. Dayal

CSIR-NGRI, Hyderabad, India

CONTENTS

1.1 INTRODUCTION

Sedimentary rocks are the weathering product of preexisting rocks. The main channels of weathering are wind and water. Based on grain size the sedimentary rocks can be divided into breccia, conglomerate, sandstone, siltstone, and shale or clay/mud rock. The basic difference between breccia and conglomerate is angular fragments in the case of a breccia and rounded fragments in the case of a conglomerate (Fig. 1.1A). The grain size of breccia and conglomerate varies between 256 and 64 mm. The cementing materials in these rocks are a matrix of fine particles. Based on grain size, the next sedimentary rock is sandstone with a grain size that varies from 0.0625 to 2.0 mm (Fig. 1.1B). Sandstone is also deposited as a weathering product of preexisting rock, but the environment of deposition is different. The weathering product is accumulated in land as well as in water bodies through wind or river as a carrier. These fine-grained transported sediments accumulate in a still environment on land or in water media. Once the thickness of sedimentation increases, pressure also increases, which leads to the precipitation of associated fine-grained material, and it behaves as a cementing material for these fine grains. It has been observed that normally the cementing material is calcium carbonate,

1

(A)   **(B)**   **(C)**

FIGURE 1.1
(A) Breccia, (B) conglomerate, and (C) shale. *From* geology.com.

both by water or air, and is later cemented by precipitation due to compaction under pressure and temperature. These fine sediments deposit on land as well as in marine conditions. Sometimes, we find red sandstone, which is due to the presence of iron oxide. Purple sandstone has been also observed, which is due to the presence of manganese oxides. Sandstone is composed of quartz and feldspar as the major minerals and muscovite, biotite, olivine, and pyroxene as accessory minerals. In some of the sandstones, heavy minerals have also been observed like rutile, zircon, and serpentine. This all depends on the weathering rock composition and transportation media.

In sedimentary rock, next in size after the sandstone is siltstone with a grain size of 0.0065 to 0.0039 mm. This particle size is between sand and shale. Normally silt occurs in the form of suspension in water media. It is the splitting of quartz grains and mixing of these grains with soil in water as a suspension. Formation of siltstone includes weathering, abrasion, and fluvial/glacial grinding. Silt rocks are composed of quartz and feldspar, but many fine-grained accessory minerals will be associated.

In sedimentary rock, the finest grained rock, with a grain size less than 0.006 mm, is claystone, mudstone, or shale. These are very fine-grained sedimentary rocks composed of very fine clay particles, including fine mica particles (Fig. 1.1C). Major minerals are quartz and calcite. Shale is part of mudstone and claystone but can be distinguished based on structure. Shales are fissile and laminated. The fine lamination is what makes it economically viable as it can store gas between the lattices in free form. Shale is observed in different colors like gray, black, green, brown, or even yellow. Like sandstone, the color of the shale indicates the presence of various minerals during their deposition. Presence of organic material provides black color, while red shale is due to presence of hematite. Presence of limonite gives them yellow color, while the presence of chlorite or illite will provide them a green color. In the formation of shale, transportation media plays an important role as these media can transport them a long distance to a quiet and calm place. At such places, these

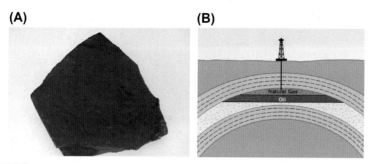

FIGURE 1.2
(A) Black shale and (B) conventional reservoirs. *From* geology.com.

fine particles accumulate, and deposition of a thick layer will become compact rock. During deposition, it is common, particularly in marine environment, for a large quantity of organic material to deposit between each layer. Once this shale sediment becomes very thick, the sediment load creates pressure and a small amount of temperature. Minor precipitation also takes place, and all these fine sediments become a fine-grained sediment rock called shale. When deposition is under reducing conditions, carbonaceous or black shale formation takes place. In carbonaceous shales, the presence of heavy minerals like uranium, vanadium, and zinc has been observed.

Shales are the mother or source rock of hydrocarbon. All types of hydrocarbon form in shale, and later after the diagenesis, these hydrocarbons migrate to reservoir rocks such as sandstone (Fig. 1.2A). Because of their very low porosity, these shales also behave as caprocks for all hydrocarbon reservoirs. In all the sedimentary basins in different formations, shale formations have been observed, but all shale formation will be carbonaceous, and economic viability is not necessary. In different times the deposition environment was different, and that is the reason you do not find oil throughout the entire geological age. During certain geological times the environment was favorable for the deposition of carbonaceous shales.

A large number of sedimentary rocks are deposited due to chemical precipitation of various chemicals and form rock salt, iron ore, chert, flint, dolomites, and limestones (Fig. 1.3). Rock salt is a product of evaporation of saline water from the ocean water or from saline lakes. Major iron formation in the world is formed from the chemicals precipitated during particular geological time eras. Iron ore is the product of chemical reactions of iron and oxygen in a marine condition. Major iron formations were deposited at 2600 Ma, 1100 Ma, and 600 Ma. All the iron ores are sedimentary rocks. Presences of ferrous oxide or ferric oxide are responsible for hematite and hematite iron ore.

(A) **(B)** **(C)**

FIGURE 1.3
(A) Rock salt, (B) iron stone, and (C) chert.

(A) **(B)**

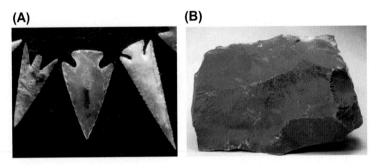

FIGURE 1.4
(A) Flint and (B) limestone. *From* geology.com.

(A) **(B)**

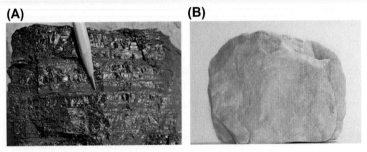

FIGURE 1.5
(A) Coal and (B) dolomite. *From* geology.com.

Chert is a silicon dioxide (SiO_2) that occurs as nodules or a concretionary mass in sedimentary rocks. It is tough material that occurs in Upper Cretaceous rocks. It is a biochemical rock and breaks with concave fracture (Fig. 1.4A). Limestone is also a sedimentary rock that forms by chemical precipitation of calcium carbonate under shallow marine conditions (Fig. 1.4B). When calcium is replaced by Mg, it is called dolomite. These rocks have great economic importance for the cement industry and other construction industries (Fig. 1.5A).

Coal is an organic sedimentary rock that forms from the accumulation and preservation of plant materials, usually in a swamp environment. Coal is a combustible rock and presently is a major rock used for fuel to produce electricity all over the world. Though there are drawbacks using coal as a major energy fuel due to large amount of carbon dioxide, sulfur dioxide, and nitrates produced, it will still is used in all the major countries all over the world.

As the price of oil has goen high, various drilling companies have tried to develop new methods for extracting the oil from a tight gas source. Earlier, it was noticed that whenever shale associated with conventional hydrocarbon was thrown as ore burden, quite often there was a generation of fire and fuming for a long time. So, it was clear the shale rocks we are throwing away are not all devoid of gas. This allows developing new techniques to extract oil from such formations that were neglected so far. With the joint operation of academia and industry, study has been conducted of such shale and how to extract the gas locked in shale layers. This study was important as there was a need to develop another source of energy in respond to the increasing hydrocarbon prices. The first time an experiment was carried out on the Barnett shale of Texas. To extract the oil, a basic need is to increase the permeability of the rock. Operators discovered that permeability of rock can be increase by fracturing under high pressure. Similar experiments were carried on various shales in Texas using various quantities of water with different pressures. The experiments were useful, and they realized that with a water pressure of 18–20K psi and by adding a small quantity of proponents, locked gas in these shales can be extracted on commercial level. This was the learning step of hydraulic fracturing of shale formation.

Earlier, only vertical drilling was carried out for conventional hydrocarbon exploration and exploitation. But later based on the structure, horizontal drilling was also carried out for the conventional hydrocarbon exploration. The earlier experience of horizontal drilling for conventional gas was useful while trying to drill the shale formation for a distance of 6–8 km. Development of horizontal drilling and high-pressure fracking in shale formation developed a new era for this source of energy. So, shale material that has been thrown away as waste material was the real source of energy. But only fracking and horizontal drilling was not enough to exploit the shale gas. The geology of the shale, mineral composition, and geochemical composition plays an important role for selecting the shale formation for hydraulic fracturing exploitation of shale gas.

Basic composition of shale is quartz and feldspar with many accessory minerals. In the field, we observe black shale, green shale, red shale, or purple shale. The carbonaceous or black color shales suggest the presence of organic material during deposition of this shale, which is done under a reducing environment.

The muddy composition of shale also defines shale as a mudstone. Some special properties like high organic content make them important resources of energy. When the organic matter is more than 1% and the deposition takes place in a reducing environment, it is called carbonaceous or black shale. These carbonaceous shales are the source rocks for the major global conventional hydrocarbons. But now with the development of shale as a reservoir and also the source rock for hydrocarbons, the global economy has changed completely. With more and more exploration and exploitation of shale gas the gas prices in the United States have come down from $8 per MMBtu to $2.5 per MMBtu. But this new source of energy has a global economic effect on conventional hydrocarbon prices, and in 6 years' time the price of oil has come down from $110 per barrel to $40 per barrel.

1.2 SHALE COMPOSITION

Shale is composed of quartz and feldspar and major minerals with many accessory minerals. The major minerals in shale are kaolinite, illite, and semectite. Other minor constituents are organic carbon, carbonate minerals, iron oxide minerals, sulfide minerals, and heavy minerals. Though shale and mudrock contain 95% organic matter, that constitutes only 1% by mass in average shale. It is not necessary that all dark color in shales are organic material. The presence of sulfide material like pyrite and deposition under a reducing environment also produce dark color in shale. If the organic material is preserved and properly heated after burial, oil and natural gas might be produced.

1.3 ENVIRONMENTS OF SHALE DEPOSITION

Shale deposition needs a quiet water column like large lakes or continental shelves. Shales are typically deposited on flood plains. They are also deposited on the continental shelf in relatively deep, quiet water. During Paleozoic and Mesozoic time, black shales were deposited in reducing environments in stagnant water columns. Fossils and animal tracks/burrows are sometimes preserved on shale bedding surfaces. Shales also contain concretions consisting of pyrite, apatite, or various carbonate minerals. The organic carbon content in black shale varies from 15 to 10%, though few types of shale contain more than 20% organic carbon. The organic matter may have generated liquid and gaseous hydrocarbon found in reservoir rocks in the subsurface into which they have migrated. The potential of gaseous hydrocarbon has been explored extensively. Study of black shale allows us to understand the past environmental history.

Mud is a chemical and mechanical weathering product of the preexisting rocks. The fine particles from this weathering are transported by wind and water and accumulate in lakes or swamps. If these accumulated sediments are not disturbed for a long time and buried, they will be transformed into a sedimentary rock known as mudstone. This is how most shale is formed.

1.4 PHYSICAL PROPERTIES OF SHALE

Shale is a very fine-grained sedimentary rock, and due to very low porosity, it does not allow any fluid pass through these spaces. Shale is a cap rock for conventional hydrocarbon reservoirs. In a tight reservoir, it is difficult to extract the oil or gas, and to overcome these problems the oil and gas industry have developed horizontal drilling and hydraulic fracturing to create artificial porosity and permeability within the rock. Petrophysics plays an important role in hydrocarbon exploration. Shale properties depend on grain size, mineralogy, porosity, and permeability. Porosity is the void space in the rock. Porosity can be divided into total porosity and effective porosity. Permeability of a shale formation depends on pore size, cementation, grading and fracturing pattern. In the case of shale gas the permeability and porosity depends on mineral composition, distribution of organic matter, amount of organic matter, and thermal maturity. The clay minerals in shale-derived soils have the ability to absorb and release large amounts of water. This change in moisture content is usually accompanied by a change in volume, which can be as much as several percent.

1.5 SHALE GAS

Shale gas contains 90% methane and 10% remaining gases like ethane, butane, and propane. Shale gas is generated and stored in situ as sorbed gas in organic matter and free gas in fractures. As such, gas shales are self-sourced reservoirs. Shales have a large amount of nonconventional oil and gas. Shale formations rich in organic material with a thickness of few hundred meters allow hydraulic fracturing. The generated gas is stored in intergranular porosity. It is also stored as adsorbed to clay particles. Based on geology and geochemistry the various shale formations will be an alternate source of energy.

The porosity of the Barnett shale was very low, and it was difficult to extract the gas or oil from this rock. The only way to extract the hydrocarbon from the shale was to change the physical properties of the shale. The permeability of shale rock can be increased by pumping a large amount of water with high pressure so that the volume of the shale can be increased and the gas stored in the lattices can be allowed to move out. Pumping a large quantity of water with some proponents and chemicals is called hydraulic fracturing (Fig. 1.5). The development of horizontal drilling and hydraulic fracturing changed the global hydrocarbon scenario in a few years.

1.6 GEOCHEMISTRY OF SHALE GAS

Geochemistry plays an important role for shale gas assessment. Geochemical analysis provides information about the total organic carbon, type of kerogen, maturity of gas, and many other parameters that help for source characterization.

Geochemical modeling of the basin helps to understand the hydrocarbon generation and retention. Pyrolysis geochemistry, carbon isotopic, and biomarker analysis of oil and gas allow us to understand the source of hydrocarbon. It is important to find the quantity and deposits of shale gas with high organic carbon. Geochemical, sedimentological, and petrophysical methods are used for the exploration of shale gas. The generated gas in intergranular spaces of shale formations is the future alternate source of energy. The geochemical method is useful for the exploration of shale gas. Shales of economic importance contain organic carbon of more than 1%, and various geological processes were responsible for the conversion of this organic matter into gas or oil.

All shale formations cannot be useful for shale gas. One of the most important properties is their hardness, as soft shale cannot sustain hydraulic fracturing. Shale should be brittle and rigid to keep the intergranular layer open so that free gas can move out (Figs. 1.6A and 1.6B). For exploration gas shale, it is important to know the total organic carbon (TOC) content. Based on TOC contents, shales samples can be selected for further studies. Rock Eval pyrolysis, gas chromatography and vitrinite reflectance analysis are useful to understand the source characteristics. The source of gas in shale is mainly of biogenic origin.

(A)

FIGURE 1.6(A)
Shale with layered structure in a folded belt.

(B)

FIGURE 1.6(B)
Shale outcrop in the field area.

It can also be divided based on maturity:

1. high thermal maturity shale,
2. low thermal maturity shales,
3. mixed lithology systems.

Methane in shale is generated from the transformation of organic material by bacterial and geochemical processes. Gas so generated gets stored by multiple mechanisms in micropores and as adsorbed on the internal surfaces. Thus shale is a combination of sorbed/micropore gas. The challenge in these accumulations is not in finding the gas but its exploitation. Thus it is more of a technological challenge that has been met by innovative hydraulic fracturing and multilateral horizontal drilling techniques.

Shale gas is alternate source of energy for the United States and other parts of the world. Though shale is the source rock for conventional and nonconventional hydrocarbon, due to its very low porosity and high permeability, conventional hydrocarbon migrates from shale to the reservoir rocks of high porosity like sandstone and limestone. Earlier, it was not possible to extract the gas from shale formations. The development of horizontal drilling and hydraulic fracturing are two main reasons for the shale exploitation. As per energy information adminstration, the United State has a reserve of 2552 trillion cubic feet (TCF) with an annual consumption of 22.8 TCF; the present natural gas resource in the United States is good for the next 110 years.

To reduce emissions, natural gas is a good source of energy. But with every industry, there are number of controversies from exploration to exploitation. With reducing reserves of conventional hydrocarbon and increasing global emission, it was necessary to find an alternate source of energy. US security for energy has greatly increased with the shale gas exploration and exploitation. While developing the shale gas, the US government took care for human health and the environment. To keep the public interest, strict guidelines have been enforced for this development. The United States in the year 2000 provided financial support and guidelines on scientific research. The initial focus was on three of the principal shale gas areas of the United States.

To set up an energy institute a good foundation base is required to form the guidelines by assessing media coverage and support from public. The guidelines were formulated with the mutual discussion of industry, academia, the community, and a team of research scientists from the Energy Institute and Tulsa University of Tulsa College in the form of a project report. To complete the objectives in a project report the team prepared major topics related to shale gas with media coverage. A state guideline as a summary in the form of a report was given to policymakers.

There was strong controversy for shale gas exploration and exploitation as the service company was avoiding giving the list of chemicals that may contaminate the ground water added in fracking fluids. Negative perceptions and political interference were sometimes responsible for the prohibition of shale gas development. Adequate protection of public health and the environment is most important. For the issues related to the environment and public health, it is important to consult the community living in the area and take their opinion as guidelines for shale gas exploration. It is very important to give correct information through the media. It is also necessary to include academic research institutes and universities as they will provide proper information, and the community will be happy with their report. There are policies and regulations for conventional oil and gas operations that will be useful for nonconventional also. Shale gas exploration and exploitation technique is different, and the current regulations for conventional oil and gas need some modification. Protecting the environment and public health should be a prime requirement to gain public support and acceptance.

For the development of well pads the activity will be construction of roads for transportation of equipment, laying of pipeline for the transportation of water for hydraulic fracturing, and construction of a reservoir with lining for different types of water and chemicals (Fig. 1.7). Such activities are important to avoid soil erosion and to protect water quality and ecological damage. During construction and operation, protection is required against leaks and spills of oil and grease and other contaminants.

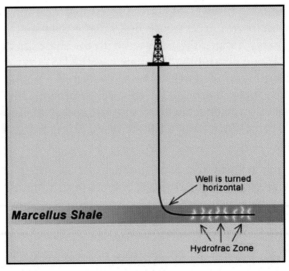

FIGURE 1.7
Unconventional oil and gas reservoir. *Source* geology.com.

The concerns over hydraulic fracturing can be summed up with a few points in question:

1. chemicals added as additives and risk to shallow water aquifers used by community,
2. impact of escaping from the shale formation and migration to aquifers,
3. any complaint from the community,
4. flowback and produced water after fracturing needing treatment before disposal,
5. well blowout or house blowout due to leakage of methane during hydraulic fracturing.

The fluid composition is 90.5% water and proppant particles, and the remaining 0.5% are various chemical additives. The additives are used to reduce friction and promote fracturing. Earlier exploration and production companies were declaring the additives as a proprietary item, but now under the present law, disclosure of chemicals used for hydraulic fracturing is mandatory. Clear information about the key chemicals used for fracturing and the impact of these chemicals for environmental toxicity is missing. Now all the chemicals used as additives for hydraulic fracturing are available in the public domain.

CHAPTER 2

Deposition and Diagenesis

A.M. Dayal

CSIR-NGRI, Hyderabad, India

CONTENTS

2.1 INTRODUCTION

Carbonaceous shale is a black color sedimentary rock commonly associated with coal formation as alternate layers of coal and carbonaceous shale. It is also associated with sandstone and limestone formations. Carbonaceous shale formations mostly consist of quartz, clays, and organic matter normally deposited in shallow marine conditions. For most of the carbonaceous shales the organic matter varies from 0.5 to 20%. Pyrolysis analysis of these shales gives us the burial history and information about maturity, type of kerogen, and depositional environment. Carbonaceous shales are the source rock for oil and gas and also for shale gas. Minor amounts of carbonate minerals are also associated with shale formations. Certain elements like trace and high field strength elements are associated with shales. Carbonaceous shales are the main resource (nonconventional hydrocarbon) of world fuel and most important reservoir of

13

Shale Gas. http://dx.doi.org/10.1016/B978-0-12-809573-7.00002-0

organic carbon in our Earth's crust in the form of oil, gas, and coal. The organic matter has generated oil and gas in carbonaceous shale, which have migrated to reservoir rock like sandstone. The potential of carbonaceous shales has been recognized in the last half decade, when the good potential of nonconventional shale gas as alternate source of fuel was realized. In a good carbonaceous shale formation the gas content will be a few trillion cubic feet.

Carbonaceous shales are distributed throughout geological time (Table 2.1). From the table, it is observed that carbonaceous shales are available in North America, South America, Asia, Australia, and Europe for all of geological time.

Such a wide distribution of carbonaceous shale over geological time suggests a good geological environment for the deposition of these shales. The organic matter in these shales has mainly marine sources, and the presence of carbonaceous shale for most of geological time suggests the evolution of life on earth from Archean to recent time. The composition of organic matter in carbonaceous shale reflects the evolutionary development of living beings during the deposition of these shales.

Depositional environments for carbonaceous shale are related to the environment such as distribution of oxygen and hydrogen sulfide in the water column. Large amounts of organic matter deplete the oxygen content in the environment, under which organic-rich sediments deposit. Hydrogen sulfide is also produced by accumulation of organic matter. Organic material is produced by photosynthesis in the photic zone mixed with some organic material from

Table 2.1 Distribution of Carbonaceous Shale Formations Over Space and Time

Age Ma	Period	Space
	Quaternary and Tertiary	Europe, Asia
65	Cretaceous	North America, South America, Europe, Asia, Africa
135	Jurassic	North America, Europe, Asia, Africa, Australia
190	Triassic	North America, Europe, Asia, and Africa
225	Permian	North America, Europe, Australia
280	Pennsylvanian	North America, Europe
320	Mississippian	North America
345	Devonian	North America, South America, Europe, and Africa
395	Silurian	North America, Europe
435	Ordovician	North America, Europe
500	Cambrian	North America, Europe, Asia
570–2500	Proterozoic and Archean	North America, Europe, Asia, Africa, Australia

terrestrial plants. The oxygen content of the water column may increase toward the bottom of the photic zone but is consumed by oxidation of settling organic matter. Hydrogen sulfide generated in the bottom sediments may diffuse to overlying water. This process creates an anoxic environment that is suitable for the deposition of carbonaceous shale.

In the sea, the oxygen content of water is controlled by circulation but also influenced by the amount of organic matter that escapes from the photic zone. The oxygen minimum is created by oxidation of organic matter. The balance between organic productivity and oxygen content may change with surface currents. The sediments deposited beneath the high-productivity zone will be rich in organic matter and suitable for the deposition of carbonaceous or black shale. In the continental shelf region the organic matter is recycled in the photic zone because of rapid settling. In this settling, sediment becomes anoxic at a very short distance below the surface because of bacterial activity. Organic productivity is controlled by nutrients, temperature, and salinity.

2.2 PHYSICAL PROPERTIES OF SHALE

Porosity is generated after deposition, and secondary porosity is very important for hydrocarbon. Secondary porosity can develop before the burial, after the burial, or as a result of other tectonic activity. Shrinkage of certain minerals creates fractures, and porosity increases. Dehydration of mud and recrystallization of minerals produces secondary porosity. Secondary porosity can be generated by dissolution of sedimentary grains or authigenic minerals. Removal of carbonate, sulfate, and feldspar creates dissolution pores.

Carbonate sediments form in shallow warm oceans by marine organisms. Normally, diagenesis takes place at the interface of the sediment, air, and fresh or sea water during different phases. Primary porosity of carbonate rock is 40–50%, and sea water sediments fill these pore spaces; this is the first stage of diagenesis. Under reducing conditions, burial diagenesis (60–70°C) generates dolomite by replacing earlier minerals at a depth of a few kilometers. Like sandstone, diagenesis in carbonates can change the reservoir properties, creating secondary porosity in carbonate. Meteoric water normally plays an important role in dissolution. With deep burial, organic material matures in source rocks and produces hydrocarbons.

2.3 ORGANIC SEDIMENTARY ROCKS

Organic sedimentary rocks are basically the product of precipitation in a marine environment. The marine microorganism also plays an important role for this precipitation. Large limestone deposits all over the world are the product of this process. Similarly, all the iron ore deposits are of marine origin and form under a certain geological condition, and that is the region you do not find iron ore deposited during all the geological events. There are specific geological time periods when iron ore was deposited. In all marine organisms, the shell formation is of dissolved calcium carbonate in sea water. Chert is deposited in sea basins as thick layers from the dead marine organisms. Similarly, coal is also formed from the deep burial of wooden material from forests. Deposition of wooden material at depth allows them to change initially into peat and with further pressure and temperature into lignite. Even today, formation of peat is continuing, but in the present environment of deposition, they cannot be converted into lignite or anthracite coal. During high weathering and heavy rainfall, this activity is more vigorous.

The hydrous minerals like rock salt, calcite, gypsum, and halite are formed during high rates of evaporation of sea water in a shallow sea. These minerals are defined as evaporites. Chert and limestone are formed by a similar process. These minerals are formed under very specific sea conditions.

2.4 DIAGENESIS: UNDERGROUND CHANGES

Sediment is accumulated by various transportation methods and deposited on land or water. As the deposition continues the thickness of accumulated sediment will increase with time. As the thickness of sediment increases with load, they are deposited deeper, and when other physical parameters start acting. The pressure and temperature will change the chemistry of deposited sediments. There could be formation of new minerals of economic importance. The depositional environment also plays an important role; as under a reducing environment, they will form different minerals than during oxidation. Conversion of limestone to dolomite and different grades of coal are examples of this process.

Sedimentary rocks are the weathering product of preexisting rock, deposit in different environments, and contain geological and environmental information about the past. The various structures on sandstone, for example, the ripple marks of different types, tell the history of deposition. Deposition in fast or slow water and wind activity will be well preserved in these ripple structures. To understand the paleo-environment, fossil impressions, mud cracks, and grain size are also useful to help understand the deposition environment of those sediments. Carbon isotopic study on these rocks gives confirmation about terrestrial or marine deposition. Strontium isotopic study on these sediments can provide information about the deposition in a lake or sea, as fresh water from a lake creates a different isotopic signature than deposition in a sea or ocean.

The basic composition of shale is quartz and calcite. The various colors of shale are due to the presence of various other materials; for example, when there is plenty of organic material, shale will be black in color, presence of chlorite will cause a slightly green color, and the presence of iron oxide will create a red color. The organic content in carbonaceous shale is around 1–10%, though for a few types of shale, up to 20% organic carbon has been observed. The presence of organic material in shale suggests the deposition under a reducing environment. In shales, clay constitutes a major part of the rock. The different minerals in clay are kaolinite, montmorillonite, and illite. In carbonaceous shale, organic matter will be more than 95%, but by weight, it is just around 1–2%. Shales are very fine-grained rocks, and they can deposit only in lakes, lagoons, and shelf regions because those are quiet environments.

Most of the black shales are deposited in marine conditions. Presently, these black shales are a major energy source along with coal, solar, and wind. Pyrolysis of these shales provides information about the type of kerogen, maturity, hydrogen, and oxygen index. The associated organic matter that generates oil and gas depends on the diagenesis. Diagenesis is the process that brings about changes in the deposited sediment under certain depositional environments. Important factors that influence the diagenesis are sedimentary or environmental factors. The sedimentary factors include particle size, fluid contents, organic matter, and mineralogy. Temperature, pressure, and chemical conditions are environmental factors. Temperature up to 200°C and pressure up to 2000 bars with change of water from fresh water to saline water play an important role in diagenesis. After deposition, sediments are buried deeply, and with sediment load, the diagenesis process starts. A large amount of organic matter is deposited near the shallow marine regions; it could be a lake, river, or ocean. When this sediment along with animal or plant organic material is buried at a deeper level due to sedimentation, the organic matter like lipids, proteins, and carbohydrates break down under temperature and pressure. The changes occur in a few hundred meters of burial of organic material, and kerogens and bitumen are formed. The kerogens formed under pressure and temperature generate oil or gas as the environment breaks down or cracks the kerogens and forms hydrocarbon. This process is called thermal cracking or catagenesis.

The various processes that change the composition and texture of sedimentary rock are called diagenesis. When the process is at a shallow level and at low temperature, it is called early diagenesis. It includes physical, chemical, and biological changes in sedimentary rock. The range of physical and chemical conditions included in diagenesis is temperature from 0 to 200°C, pressure from 1 to 2000 bars, and water salinities from fresh water to concentrated brines. In fact the range of diagenetic environments is potentially large, and diagenesis can occur in any depositional or post-depositional setting in which a sediment or rock may be placed by sedimentary or tectonic processes. This includes deep burial processes but excludes more extensive high-temperature or high-pressure metamorphic processes.

2.5 DIAGENESIS

All the changes that occur in sedimentary rock after deposition can be classified under the term diagenesis. In metamorphism, pressure and temperature also play an important role, but metamorphism is entirely different than diagenesis. Diagenesis includes compaction, deformation, dissolution, recrystallization, and micro bacterial activity. In the case of metamorphism, tectonics also plays an important role, and various folds and fault are a part of metamorphism. Compaction and lithification are important process of diagenesis and responsible for rock hardness.

During initial deposition of sedimentary material the porosity of the sediment will be very high, as there is no compaction factor. In such rock, 30–70% porosity has been observed. Porosity of the rock also depends on the grain size, as the coarser grains will have higher porosity than the fine-grained sediment because the air space in coarse grain is quite high in comparison. In clay or mud deposits, initially, the porosity will be high because there are water molecules associated with them, but as the water dries up, the porosity changes, and later, porosity will be very low. Porosity of carbonate rock depends on the type of sediments. There could be well-sorted sediment, or it may be a mixture of coarse and fine grains, which indirectly reflects the deposition environment. Weathering product is transported and gets accumulated at one place; this is one activity. Now, as this accumulation process continues, there will be a compaction phenomenon that comes into the process. With compaction, accumulated material is buried at depth, and the depth will change as more and more sediment is deposited.

After a certain depth, the diagenesis process takes over, and there will be various physical, chemical, and biological activities that will change the composition of sedimentary rock. The various physical parameters of the rock also change, such as porosity and permeability. These processes will modify the original chemical composition. Chemical diagenesis is very important for the economics of geology as it affects the siliciclastic and biogenic materials. This includes minerals reactions, their solubility, and pore fluid interaction and biochemical reactions. For chemical diagenesis the pore fluid plays an important role, as its chemistry and temperature are important for the depletion of trace elements (Einsele, 1965). Diagenesis of sandstone and limestone is important because they are the reservoir rocks for the petroleum system. Sediments deposited under anoxic conditions are rich in organic matter and form early diagenetic minerals. Semectite forms illite between 60–100°C, and kaolinite converts to illite and quartz at 120–150°C. Most of these reactions release water and change the composition of pore fluids.

Sedimentary factors include compaction of particles under pressure, cementation of particles by precipitation, recrystallization, and replacement of particles. These transformations can change the porosity and permeability, flow rate, and reservoir volume. Porosity and permeability are controlled by sedimentary conditions at the time of deposition. The environment of deposition is responsible for the diagenesis. Deposition of carbonates is controlled by marine biological activity in which light, nutrients, salinity, temperature, and turbidity plays an important role.

The diagenesis process could be early diagenesis, middle diagenesis, or late diagenesis. An early diagenesis takes place at shallow depth up to a few hundred meters and temperature below 140°C. In this case, organic material loses pore water, resulting in the formation of lignite or subbituminous coal. In early diagenesis, chemical, physical, and biological processes influence the organic matter by bioturbation and bio-irrigation. There is ion exchange on the surface and intra-sheet layers of clay minerals, which results in the formation of new minerals. There is partial or complete dissolution of minerals. Bacterial degradation of organic material includes aerobic respiration in the presence of oxygen, and denitrification, manganese, iron, and sulfate reduction occurs. Physical effects are responsible for the burial and overburden of sediment and compaction of sediment.

Pore fluids, which include meteoric water, connate water, and juvenile water, play an important role in diagenesis. Sources of these waters are rain or snow for meteoric water, water trapped in sediments during deposition for connate water, and water of magmatic origin for juvenile water.

If one observes the sedimentary profile, there is an uppermost zone, where pores are open to the atmosphere; it is also called the vadose zone. The next zone is the saturated water zone, which is separated by ground water from the uppermost zone. Aquifers occur in fractured sedimentary rock with high porosity and have the capacity to accommodate a large quantity of water in the porous space. There are many sedimentary rocks with high porosity/permeability, and these rocks also accommodate a large quantity of water or liquid material. These are sandstone, gravel, and conglomerate, also termed reservoir rock. The sedimentary rock with low porosity and permeability, such as shale or clay rock, is very poor as a reservoir rock and is called an aquiclude.

The chemical composition of connate water changes, as it is the product of the chemical reaction of sediment and fluid at the time of deposition. The salinity, temperature, chemistry, and carbon isotopic ratio of connate is quite different from meteoric water. Compared to fresh water, connate water is denser due to the presence of various salts. At depth, the meteoric water is displaced by connate water. In weathered zones, meteoric water can increase the porosity of sandstone and limestone because water is the universal solvent.

2.6 DIAGENESIS OF FINE-GRAINED SEDIMENTS

Fine-grained sediment normally deposits in marine environments, and diagenesis activity for these sediments is very little. As these sediments deposit in a shallow marine environment the presence of oxygen slows down the diagenesis activity. In a lake or swamp area, dehydration takes a very long time, which allows the organic matter to be oxidized. Hematite is dehydrated part of limonite, which is precipitated as iron sulfate. But in the subsurface the oxygen is almost absent, and under a reducing environment, sulfate bacteria convert the ferrous iron to mineral pyrite (ferrous iron sulfide).

$$Fe(OH)_2 + 2S = FeS_2 + H_2O$$

The reaction of sulfur with organic matter will form hydrogen sulfide.

$$SO_4 + 2CH_2O = 2HCO_3 + H_2S$$

Fermentation of organic matter through bacteria also takes place at shallow depth, and it generates water, carbon dioxide, and methane. As the pH of the pore fluids increases, it will allow precipitation of carbonate.

The mineral composition of freshly deposited clay is illite, kaolinite, chlorite, and semectite in different proportions. There are significant changes to these clay minerals under diagenesis, and temperature plays an important role. At further depth with change of temperature the common mineral changes are semectite to illite and kaolinite to illite and chlorite. But during metamorphism the illite changes to fine-grained mica as sericite.

2.7 THE ROLE OF DIAGENESIS IN HYDROCARBON GENERATION

During sedimentation, organic matter is also deposited with fine sediments. As the thickness of the deposited sedimentation increases with depth, the organic matter under pressure and temperature converts into kerogen and bitumen. These kerogens break down into hydrocarbon by various chemical processes known as cracking or catagenesis. Based on the temperature and pressure action on the organic matter, the result will be early, middle, or late diagenesis. In early diagenesis, formation of various types of coal takes place. In early diagenesis or eodiagenesis, hydrocarbons cannot be generated. In middle or mesodiagenesis, due to dehydration, genesis of gas/oil occurs, and bituminous coal formation also takes place. In the case of late diagenesis or telodiagenesis, there is cracking of kerogen, and dry gas or anthracite coal formation takes place.

2.8 VARIOUS PHASES OF DIAGENESIS

The various phases of diagenesis are very complex and take much geological time. Temperature plays an important role for montmorillonite degradation. The increased geothermal gradient is responsible for illite to montmorillonite transformation at depth, probably due to high temperature. Temperature plays an important role during catagenesis and diagenesis processes. In the case of shale gas the formation of methane and its relation with pore pressure, diagenesis, or catagenesis is not well understood. The transformation of various minerals in shale gas needs the active role of water. In the case of low thermal gradient, activity of pore pressure will be also low. During mesodiagenesis, clay minerals get dehydrated, and genesis of oil takes place. In the case of telodiagenesis, the temperature is high, which allows cracking that results in the formation of dry gas and poor quality anthracite coal. Abnormally high formation pressure is related with stress related mechanisms, increase in fluid volume (probably due to increase in temperature), movement of fluid, and hydrocarbon buoyancy. Mineral transformation is also responsible for high pressure in shale formation.

Carbonaceous shale formation of a marine and nonmarine nature is a major source of energy. Due to low porosity and high permeability of the carbonaceous shales the oil and gas migrate to the reservoir rocks like sandstone. Study of marine carbonaceous shales also provides information about the evolutionary history of the living beings. The amount of metal contents in shale provides information about the depositional environment in shelf regions. The carbonaceous shales have deposited in different geological settings in the shelf region.

Sedimentation rates and organic productivity are the important parameters for controlling the deposition of the sediments as carbonaceous shales. Oxygen plays an important role for the formation of carbonaceous shales. A large quantity of organic material depletes the oxygen and favors the formation of carbonaceous shales. Hydrogen sulfide also helps in the formation of carbonaceous shale. Under restricted circulation of oxygen and water, organic-rich sediments can accumulate, even with small organic productivity. Basically, organic material is produced by photosynthesis in the photic zone. A large quantity of oxygen is required for the decomposition of organic material. In the ocean, oxygen content is controlled by circulation of water and organic material coming from the photic zone. In the continental shelf region, there is very little water between the photic zone and sea bottom; this is responsible for the settling of the organic matter to the bottom. Availability of nutrients, temperature of the water, and salinity are responsible for the organic productivity.

Pyrocatalytic formation begins at 50°C, but it is a minor amount; major formation is produced at higher temperatures. Pyrocatalytic methane will have a higher $\delta^{13}C$ value than biogenic methane. For the characterization of the hydrocarbon zone and its economic importance, study of organic composition, vitrinite reflectance, and color of alteration is necessary. It is also necessary to understand the conversion of montmorillonite to illite, as this allows the release of water as a by-product that helps the migration of hydrocarbon in the reservoir zone. For the diagenesis, deep burial of organic material and time is responsible for the pressure and temperature.

Table 2.2 Zone of Diagenesis With Depth, Temperature and Porosity

Depth (km)	Temperature (centigrade)	Porosity (%)	Diagenetic Zones and Products
0.01	0	80	Sulfate reduction Pyrite carbonate (^{12}C enriched) Fermentation Methane carbonates (^{13}C enriched)
1	30	30	Decarboxylation Formation of siderite
2	70	20	Hydrocarbon formation Pyrocatalytic methane Montmorillonite to illite
3	190	10	Metamorphism Graphite, recrystallization

Important factors responsible for the diagenesis are particle size, fluid content, organic contents, and mineralogical composition. Pressure, temperature, and chemical reaction are important parameters for the porosity and permeability of the sedimentary rock, which is controlled by the process and time of deposition during the deposition of sedimentary rock. Depositional environment will be different for sandstone, shale, and limestone. Deposition of sandstone is related with low or high sea level. But for limestone, the depositional environment is related with biogenic activity.

According to Fairbridge (1966), there are three distinct phases of diagenesis. These are syndiagenesis, anadiagenesis, and epidiagenesis. Water, being a universal solvent, plays an important role in diagenesis. Clays are important for the diagenesis process and behave as a cementing agent for different minerals. Allogenic clays are a dispersed matrix carried downward by migrating pore water. Allogenic clays are also produced through ingestion and excretion by organisms. Authigenic clay forms within the sandstone after the burial. Pore water and rock chemistry plays an important role for the formation of authigenic clay. Authigenic clay in sandstone can be observed as a coating to the grains, pore filling, and filling the fractures.

Sandstone is an assemblage of various minerals with particle size of 0.0625–2 mm diameter, and it is composed of mainly quartz, feldspar, and other minerals. Sandstone forms under a certain temperature, pressure, pH, and oxidation state. The diagenetic process starts between the depositional medium and layers of sediments. Deep burial reduces porosity and increases permeability, while cementation reduces the pore spaces. In the case of ductile grains, they will change their shape under pressure. Pressure dissolution reduces porosity and volume and also forms new minerals. Sometimes, silicate grain is replaced by carbonate mineral and also changes the porosity. Some minerals have the tendency to change to other minerals. Cementing material in sandstone is a precipitated material from pore fluids. Some of the common cementing materials in sandstone are quartz, calcite, dolomite, feldspar, gypsum, anhydrite, zeolite, barites, and clay minerals like illite, kaolinite, and chlorite.

2.9 ZONE OF FERMENTATION

In the zone of fermentation, methane is produced by bacterial reduction of carbon dioxide. The fermentation production of biogenic methane is also responsible for the large fractionation of carbon isotopes. Biogenic methane is expected with very light carbon isotopic composition. With increasing temperature and depth of burial, the sediment reaches to the zone of decarboxylation, where organic matter begins to decompose by chemical process. Movement of organic matter from the lower part of decarboxylation to the upper part is the zone of hydrocarbon formation. Pyrocatalytic formation of methane begins at 50°C, but a large amount of methane is produce at much higher temperature. The carbon isotopic signature for pyrocatalytic methane is higher than that of biogenic methane. Burial depths and thermal gradients are responsible for the transformation of various minerals.

References

Einsele, G., 1965. Sedimentary Basin: Evolution Facies and Sedimentary Budget. Edtited published by. Springer Verlag, Berlin. 781 p.

Fairbridge, R.W., 1966. Diagenetic phases: abstract. AAPG Bull. 50 (3), 612–613.

CHAPTER 3

Organic Matter in Gas Shales: Origin, Evolution, and Characterization

D. Mani, M.S. Kalpana, D.J. Patil, A.M. Dayal

CSIR-NGRI, Hyderabad, India

CONTENTS

Shale Gas. http://dx.doi.org/10.1016/B978-0-12-809573-7.00003-2

3.1 INTRODUCTION

Sedimentary organic matter when exposed to sufficient temperature and pressure over geological time in the subsurface cracks gives oil and gas. The gaseous hydrocarbons in gas shales are generated and stored in the organic-rich, fine-grained matrix of shale rocks. The organic matter is derived from the cells or tissues of once living organisms and requires favorable environmental conditions for its production in large quantities. Depositional conditions accumulate the organic matter in sediments and its preservation and evolution into hydrocarbons is governed by the post-depositional environments. The production, deposition, preservation, and maturation of organic matter define the quality and quantity of shale gas plays (Fig. 3.1, Mani et al., 2015a).

Photosynthesis forms the basis of mass production of marine or terrestrial organic matter on earth. It is a complex process in which plants, algae, and certain type bacteria (e.g., cyanobacteria, purple sulfur bacteria) convert the energy of sunlight into chemical energy for the synthesis of organic compounds. The process takes place in the plant cell organelle, chloroplast, where the green colored pigment–chlorophyll–catalyzes the reaction.

$$6CO_2 + 12H_2O \xrightarrow{h \cdot v} C_6H_{12}O_6 \text{ (Glucose)} + 6O_2 + 6H_2O \tag{3.i}$$

Glucose, the six carbon monomer unit, formed in reaction acts as a building block for large quantities of organic matter fixed by the autotrophic organisms in the form of polysaccharides such as cellulose and starch and is also used in synthesis of proteins and lipids. Nearly 75 million petagrams (Pg) of carbon are distributed within various reservoirs of the Earth's crust, biosphere, and ocean (Petsch, 2014). The majority of this carbon is concentrated in the crustal rocks,

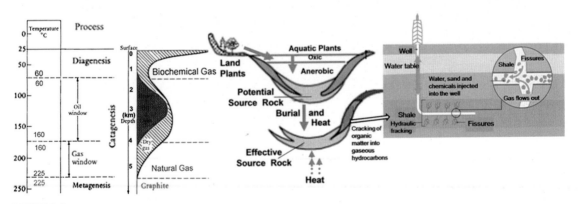

FIGURE 3.1

Production, deposition, preservation, and maturation of organic matter in shale gas plays. *Modified after Mani, D., Patil, D.J., Dayal, A.M., 2015a. Organic properties and hydrocarbon generation potential of shales from few sedimentary basins of India, petroleum geosciences: indian context. Springer Ser. Geol. 99–126.*

primarily the sedimentary rocks, in the form of inorganic carbonates (~60 million Pg) and organic carbon (~15 million Pg) (Petsch, 2014). The dynamic equilibrium between the two forms is manifested in the global carbon cycling. The yearly production of organic matter by terrestrial and marine biosphere (i.e., the primary production) is 105 Pg per year (Petsch, 2014). A fraction of less than 0.1% of the organic matter gets preserved in the sediments, depending upon the geological and geochemical conditions of sedimentation.

Organic matter is composed of carbohydrates, proteins, and lipids, of which the lipids are the most resistant to the sedimentary degradation processes. The incorporation ratio, defined as the organic matter that gets incorporated into sediments versus the organic matter photosynthesized by the same of surface unit, varies upon the geological environment (Bordenave, 1993). Sediments having high organic matter (>1%) are located on edge of continents and islands, in continental shelves, and those parts of abyssal areas and arc-related trenches where organic matter is transported together with sediments by turbiditic current (Bordenave, 1993). Organic content of marine sediments is low (<1%); however, marine areas such as offshore Peru, the Pacific offshore of the USA, Coast of Africa, Sea of Oman, and Bay of Bengal have organic matter >1%. Inland seas such as Black Sea, Baltic Sea, and northern part of the Barents Sea have organically rich sediments, and the incorporation ratio is also high. Productivity is relatively high in the coastal areas, where upwelling zones prevail.

The quantity and quality of organic matter preserved during diagenesis of sediment ultimately determines the hydrocarbon generative capacity of the rock. Diverse factors influence the preservation of organic matter during sedimentation and burial. Higher organic productivity, nutrient supply (brought by rivers, upwelling currents, and water mixing), low oxygen content of associated water column and sediments (less than 0.2 mL/L water), restricted water circulation, lack of bioturbation, very fine-grained sediment particles (<2 μm), and an optimum sedimentation rate favor the preservation of organic matter in the sediments.

A small amount (10–20%) of petroleum is formed from the hydrocarbons, mainly the lipid content, which has been synthesized naturally by the earlier living organisms (Hunt, 1996; Mani et al., 2015a). Its main constituents are the carbon homologous (≥C15+) compounds that have identifiable biological and chemical structure or biomarkers (Mani et al., 2015a). These molecules get embedded in sediments with minimal or slight structural changes during the digenetic processes related with sedimentation (Hunt, 1996; Mani et al., 2015a). Biomarkers are soluble in organic solvents and constitute the bitumen part of organic matter (Hunt, 1996; Mani et al., 2015a). The second major source, which constitutes about 80–90% of hydrocarbons, involves the conversion of the proteins, lipids, and carbohydrates of living beings into organic matter of sedimentary rocks, and it is called kerogen (Hunt, 1996;

Mani et al., 2015a). Consecutive burial of organic matter at different depths in the subsurface leads to various chemical and physical transformations in it and ultimately leads to the formation of liquid and gaseous hydrocarbons, through the stages of diagenesis, catagenesis, and metagenesis (Hunt, 1996; Mani et al., 2015a). Diagenesis occurs in recent sediments at shallow depths where the temperatures are less than 60°C. The bio-polymers like carbo-hydrates and proteins evolve into geo-polymers, or kerogen. Further, with time, sedimentation leads to burial of previously deposited beds and exposure of sediments to subsurface conditions of increasing temperature (T~50–150°C) and pressure (P~300–1000 bars) (Mani et al., 2015a). Overburdened sedimentary column and tectonic activities govern the subsurface P–T conditions. Thermal degradation of kerogen results in the formation of petroleum, condensates, and wet gases. At still higher temperatures (150–200°C), metagenesis processes operate, and the organic matter is cracked to dry gas. Metamorphism is the last stage of evolution of sediments in the subsurface (Hunt, 1996; Mani et al., 2015a).

3.2 ORGANIC MATTER IN GAS SHALES: QUALITY

The main component of the shale gas is methane, which comprises of about 70–90%, along with lesser amounts of light hydrocarbons such as ethane, butane, carbon dioxide, oxygen, nitrogen, hydrogen sulfide, radon, and other rare gases. The quantity, quality, and thermal maturity of organic matter accounts for the gas generation capacity of shales. The higher the organic content of rocks is, the greater is the potential for hydrocarbon generation. The type or quality of organic matter is dependent upon the source of organic components and its environment of preservation, thus resulting in different kerogen types with varying capacity for oil and gas generation. Microscopic examination of organic facies and programmed pyrolysis of organic matter provide necessary parameters such as TOC content, type of kerogen, and its thermal maturity for the qualitative and quantitative evaluation of kerogen.

3.2.1 Total Organic Carbon Content

The quantity of organic carbon is relative to the total organic carbon (TOC) content of rocks. Thus, TOC is the concentration of organic matter in the rocks and is expressed by the weight percent of organic carbon. A value of about 0.5% TOC by weight percent is considered a minimum or threshold for an effective source rock. For shale gas reservoirs, in general, values of about 2% are considered a minimum and may exceed 10–12% (Table 3.1). Typical TOC values for the Barnett and Marcellus shales are 2–6% and 2–10%, respectively. Thus, in terms of TOC content, Marcellus is organically richer.

Table 3.1 Total Organic Carbon (TOC) Content and the Hydrocarbon Generative Potential of Source Rocks

Total Organic Carbon Content Weight %	Hydrocarbon Generation Potential
<0.5	Very poor
0.5–1	Poor
1–2	Fair
2–4	Good
4–12	Very good
>12	Excellent

Table 3.2 Hydrocarbon Potential of Different Kerogen Types Based on the Depositional Environment

Depositional Environment	Kerogen Type	Maceral Type	Origin	Hydrocarbon Potential
Aquatic	I	alginite	Algal bodies	+
		Amorphous organic matter	Structureless debris of algal bodies	oil
	II		Structureless, planktonic material, primarily of marine origin	oil
		exinite	Skins of spores and pollen, cuticle of leaves, herbaceous plants	
Terrestrial	III	Vitrinite	Fibrous and woody plant fragments and structureless colloidal humic matter	Gas and some oil
	IV	Inertinite	Oxidized, recycled woody debris	None −

Modified after Tissot, B.P., and Welte, D.H., 1984. Petroleum Formation and Occurrence. Springer-Verlag, New York.

3.2.2 Organic Facies and Kerogen Type

A majority of organic matter in sedimentary rocks occurs in the macromolecular form of kerogen, which is the precursor for generation of hydrocarbon. Its concentration, composition, and properties in the rocks are vital for resource assessment. These properties help in understanding the origin of petroleum and natural gas, thermal history of sedimentary basin, and depositional environment of source rock (Table 3.2). Investigation of the kerogen properties is also critical to further the understanding of the manner in which gas shales store, retain, and release natural gas. The potential to generate gas in shale rocks depends mainly upon the organic facies. It has been defined as "a mappable subdivision of a designated stratigraphic unit, distinguished from other adjacent subdivisions on the basis of character of its organic components, without regard to the inorganic aspects of the sediment" (Jones and Demaison, 1982; Jones, 1983; Mani et al., 2015a). The organic facies are composed of microscopic organic materials or macerals, which are derived from terrestrial, marine, and lacustrine plant remains. These are a function of the parent source material, decomposition during early diagenesis, and degree of thermal exposure in the subsurface, which leads to the evolution or maturation in different forms. Macerals behave differently in their physico-optical properties and have been classified into three groups: liptinite, inertinite, and huminite or vitrinite. The maceral group—liptinite—comprises the optically distinctive parts of plants like spores, suberine cuticles, etc., and various products due to degradation and maturation processes. These macerals have the highest hydrogen content and consist of aliphatic compounds. Upon thermal transformation, these evolve into hydrocarbons, largely oil (Tissot and Welte, 1984). Inertinite is derived from strongly altered and degraded plant material and constitutes products that are derived from aromatization and condensation reactions (ICCP, 2001). The huminite or vitrinite maceral groups are derived from cellulose and lignin parts present in plants, tannins of woody tissues, and colloidal humic gels (ICCP, 2001; Sýkorová et al., 2005). Their formation involves processes such as humification and biochemical and geochemical gelification (Sýkorová et al., 2005). These macerals are represented by aromatic structures in the low-rank coals, and increasing coal ranks leads to a notable increase in the aromaticity and condensation. Upon exposure to higher temperatures in the subsurface, these predominantly produce gaseous hydrocarbons.

The reflectance of vitrinite maceral, expressed as %Ro, is used as a proxy for the level of organic maturity (LOM). Thermally immature source rocks do not have a pronounced effect of heat (<0.6%Ro) (Mani et al., 2015a). Ro range between 0.60 and 0.78 usually represents oil-prone intervals, and the values of Ro>0.78 indicate gas-prone kerogen (Mani et al., 2015a). High values are indicative of "sweet spots" for the gas shale wells. Thermally mature organic matter generates oil (0.6–1.35%Ro), whereas the post-mature organic matter is in wet and dry gas zones (Fig. 3.2)(Mani et al., 2015a). The vitrinite reflectance of Barnett shale, marking the beginning of dry gas (%Ro = 1.2) is considerably lower than other shales like Marcellus (%Ro = 1.6).

The dominance of oil and/or gas in the product as kerogen cracks through time under influence of heat in the subsurface is determined by the type of organic matter and level of thermal maturity. Depending upon the source and input of organic matter in a particular depositional environment, the elemental content such as carbon (C), hydrogen (H), and oxygen (O) of kerogens differs, resulting its classification into Type I, II, IIS, III, and IV (Van Krevelen, 1961; Mani et al., 2015a). Type I kerogen is generated mainly from the lacustrine type of depositional settings and marine environments and is composed mainly of

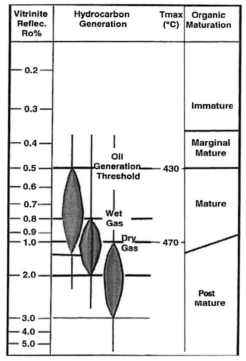

FIGURE 3.2

Vitrinite reflectance and level of organic maturation (LOM) in source rocks. *Modified after Tissot, B.P., and Welte, D.H., 1984. Petroleum Formation and Occurrence. Springer-Verlag, New York.*

liptinite maceral (Hunt, 1996; Tissot and Welte, 1984). It is derived from algae, plankton, and other types of organic matter that has been thoroughly reworked by bacteria or microorganisms dwelling in the sediment. It consists of long aliphatic chains and has higher hydrogen with low oxygen content. Primarily, it is oil prone in nature, but with increasing thermal maturation, it can also produce gas. Type II kerogen is normally generated in the reducing and oxygen-deficient environments of fairly deep marine settings (Tissot and Welte, 1984). It is derived from the planktonic remains that have been reworked and altered by bacteria. Type II kerogen consists of aliphatic chains, but they have more aromatic and naphthenic structures compared to Type I. It is rich in hydrogen and low in carbon, and it can generate oil or gas with increasing heating and maturation (Tissot and Welte, 1984). Type II-S kerogen is formed in certain depositional environments that promote the increased incorporation of sulfur compounds in the kerogen. The generation of oil in Type II-S kerogen starts much earlier, due to kinetic reactions involving sulfur-containing compounds. Type III kerogen is derived chiefly from terrigenous plant debris, which get deposited in shallow to deep marine or nonmarine environments. It contains minor aliphatic chains with mostly condensed poly-aromatics and oxygenated functional groups (Mani et al., 2015a). Type III kerogen contains primarily the vitrinite maceral and has low hydrogen and high oxygen content and is dry gas prone in nature. Coals, in general, consist of Type III kerogens. Type IV kerogen originates mainly from residual organic matter and is present in older sediments reworked after erosion. The kerogen content is strongly altered by weathering, combustion, and biological oxidation. This type of kerogen has high residual or inert type of carbon content with very low hydrogen, thus having almost no potential for oil or gas (Hunt, 1996; Tissot and Welte, 1984).

The amount of hydrogen and oxygen in a kerogen is also determined by the Rock Eval pyrolysis. The parameter HI (hydrogen index) refers to the hydrogen richness of the source rock, and OI (oxygen index) measures the oxygen richness. These parameters are used to assess the quality of petroleum source rocks. Organic-rich shales deposited in reducing, anoxic marine environments have higher HI values and low OI values for which modified a van Krevelan diagram (Van Krevelan, 1961; Tissot and Welte, 1984; Mishra et al., 2015) is used to evaluate the quality of kerogen based on these indices. Rock samples with HI value < 50 mg HC/g TOC (mg hydrocarbon per gram of TOC) contain Type IV kerogen, which suggests no potential of it to generate oil and gas (Mishra et al., 2015). HI values that range between 50 and 200 mg HC/g TOC suggest Type III kerogen with gas generation potential, whereas those between 200 and 300 mg HC/g TOC suggest Type- II/III kerogen (mixed oil and gas) (Mishra et al., 2015). An HI value between 300 and 600 mg HC/g TOC suggests oil-prone Type- II kerogen and >600 mg HC/g TOC suggests Type I kerogen with generation of oil (Peters and Cassa, 1994; Mishra et al., 2015; Mani et al., 2015a,b).

Mixed Type II and Type III kerogen constitutes most of the organic matter in currently producing shales. Barnett kerogen is characterized by Type II organic matter with a minor contribution from Type III, whereas Marcellus kerogen contains a slightly greater mixture of Type III. As Type II kerogen generates a greater quantity of hydrocarbons and at a lower temperature, the conversion of Barnett organic matter to natural gas was probably a more effective process compared to that of Marcellus. Even though gas-prone source rocks can generate large quantities of gas at higher maturity, they may not account for the major source of gas in the subsurface. As oil generation reaches near completion, the oil-prone kerogen may still have significant capacity and potential for generating gaseous hydrocarbons. About 20% and 30% of the oil and/or bitumen generated by oil-prone kerogen may be retained in the source rock, which eventually cracks to form a significant amount of gas. The late gas generation during cracking of residual oil or bitumen that occurs in oil-prone source rocks can contribute more toward generation of gas than gas-prone source rocks.

3.2.3 Thermal Maturity of Kerogen

The amount of heat experienced by the source rock in the subsurface is critical for the generation of hydrocarbons. Thermal maturity is the extent of temperature–time driven reactions, which are responsible for the conversion of sedimentary organic matter to oil and gas (Mani et al., 2015a). Vitrinite reflectance (Ro%) and Rock Eval pyrolysis temperature (Tmax) are popularly used parameters to assess the thermal maturity of kerogen. Thermally immature source rocks do not have a pronounced effect of temperature (<0.6%Ro) (Mani et al., 2015a). Thermally mature organic matter generates oil (0.6–1.35%Ro), whereas the post-mature organic matter is in wet and dry gas zones (Tissot and Welte, 1984; Mani et al., 2015a). The Tmax values <435°C indicate an immature stage for generation of hydrocarbons; the temperature range between 435 and 465°C suggests a mature stage, and those >465°C show a post-mature stage, suitable for generation of gas (Tissot and Welte, 1984; Hunt, 1996; Mani et al., 2015a).

3.2.4 Depositional Environments

Many prolific shale gas reservoirs of today are over-mature, oil-prone source rocks. These reservoirs evolved from organically rich fine sediments or mud deposited in marine, lacustrine, or swampy environments through burial and heating over geological times. Nearly all shale gas plays evolve from oil-prone marine organic matter, which has been produced autochthonously due to large-scale algal blooms occurring in surface waters of oceans and seas. Sea currents and river mouths are the main sources of organic runoff from land because the shale rocks which have the highest organic matter content are developed in shallow shelf seas close to the widespread delta systems of rivers.

The ideal condition for source rock formation is the density stratification and upwelling cases in which there is a superposition of oxygenated nutrient-rich eutrophic zone and very low-energy, oxygen-free water at depth. One of the most prominent gas plays in USA, the Barnett of Fort Worth basin, Texas, was deposited in a deeper water foreland basin having poor circulation with the open ocean during most of the basin's development and depositional history (Loucks and Ruppel, 2007). The bottom waters were euxinic, which led to preserving organic matter and resulted in rich source rock (Loucks and Ruppel, 2007). Abundant framboidal pyrites are also present, which are formed in such a kind of depositional setting (Loucks and Ruppel, 2007). High-salinity conditions and water density stratification prevailed during deposition of Woodford shale, south-eastern Oklahoma (Romero and Philip, 2012). For the rocks that have a highly oxygenated depositional and early diagenetic environment, the organic matter content is generally low, and the kerogen has negligible generative capacity for hydrocarbons (Mani et al., 2015a). An anoxic environment can result in the deposition of organic-rich, fine-grained sediments that can develop into excellent potential source rock (Mani et al., 2015a).

3.2.4.1 Molecular Signatures

A significant constituent of sedimentary organic matter, the biomarkers, are partial fossil records (Moldowan et al., 1985), which provide considerable information about the contributions from different biota during the geological past and are useful in several aspects of petroleum exploration. The usefulness of biomarkers as an indicator of depositional environment stems from the fact that these naturally derived molecular fossil compounds are associated with certain assemblage of plants or organisms that grow in a specific type of environment (Seifert and Moldowan, 1986). Due to their low susceptibility toward microbial and diagenetic degradation, the biomarkers, such as alkanes including steranes and hopanes, can record many aspects of depositional history and sources of organic contribution according to the occurrence of terrestrially or marine-derived organic matter in the rock or a bacterial source for it (Peters et al., 2005). The source and maturity of organic matter affects the odd and even carbon number predominance in a hydrocarbon compound (Tissot and Welte, 1984). Carbon preference index (CPI) is the ratio obtained by dividing the sum of odd carbon-numbered alkanes to the sum of even carbon-numbered alkanes (Didyk et al., 1978). The pristine-phytane ratios (Pr/Ph) in a particular sedimentary rock are used as an indicator for understanding the different types of depositional environments (Didyk et al., 1978). Samples with high Pr/Ph ratios (<3) indicate inputs of land-derived organic matter of higher plants, which have influence of oxidation/weathering prior to preservation. Low Pr/Ph values (<2) indicate aquatic depositional environments such as marine, fresh, and brackish water having reducing conditions (Lijmbach, 1975). The intermediate values (2–4) indicate a fluvio-marine and coastal swampy environment (Lijmbach, 1975). High values (up to 10) are related to peat swamp depositional environments with oxidizing conditions (Lijmbach, 1975). Aromatic markers such as polycyclic aromatic hydrocarbons (PAHs) are ubiquitous components of both marine, lacustrine, and terrestrial sediments, and the relative abundance and ratios of PAHs reflect the thermal maturity of sedimentary organic matter (Alexander et al., 1985).

3.2.4.2 Stable Carbon Isotopic Signatures

Stable carbon isotope ratios are generally used to identify origin of diverse potential sources of the hydrocarbon gases, particularly methane (Schoell, 1983; Mani et al., 2015b). The depositional environment of shales, being marine or nonmarine, has a direct influence on the type and amount of organic matter that they contain. The marine and continental (terrestrial and freshwater) plants show different $\delta^{13}C$ signatures due to the difference in the isotopic composition of carbon sources (Mani et al., 2015b). As a result, the gas systems originating from a respective source of organic matter have been classified into two distinct types, namely biogenic (or microbial) and thermogenic. There can also be mixtures of the two gas types (Jarvie et al., 2007; Mani et al., 2015b).

Occurrence of thermogenic gas in the subsurface establishes the possibility or presence of a source rock and provides information on the elements of the petroleum or gas system, which otherwise lacks in biogenically produced gases (Mani et al., 2015b). Because the stable carbon isotope ratios are susceptible toward fractionation due to migration, thermal maturation of hydrocarbons, and bacterial oxidation, care should be taken that the shales are not contaminated by weathering processes (Mani et al., 2011, 2015b). The thermally desorbed hydrocarbon gases upon pyrolysis of shales reflect the pristine signature of gaseous components derived from the thermal cracking of kerogen (Mani et al., 2015b).

3.3 ORGANIC MATTER IN GAS SHALES: QUANTITATIVE APPROACH

The quantitative approach of petroleum evaluation in a sedimentary basin takes into account the amount of oil and gas generated by primary cracking of kerogen when temperature increases with time (Tissot and Welt, 1984; Mani et al., 2015a). The kerogen degradation is described by a series of parallel chemical reactions, each of which obey first-order kinetics, and they are characterized by Arrhenius Law:

$$k\,(\mathrm{T}) = A * \exp\left(-E/\,(RT)\right) \tag{3.1}$$

Where k is the reaction rate parameter dependent on absolute temperature (T, in Kelvin), E is the activation energy of reaction (J/mol), and R is the molar gas constant (J/mol/K); A is the pre-exponential frequency factor/Arrhenius constant (sec^{-1}) (Tissot and Welt, 1984; Mani et al., 2015a).

Eq. (3.1) is expressed as follows:

$$dXi/dt = A \exp\left(-Ei/RT\right).Xi \tag{3.2}$$

dXi/dt is the hydrocarbon generation rate, and Xi is the residual petroleum potential of the organic matter involved in reaction i. E is the activation energy related to reaction i (Tissot and Espitalie, 1975; Tissot et al., 1987; Ungerer et al., 1986; Braun and Burnham, 1987; Ungerer and Pelet, 1987; Tissot and Welte, 1984; Mani et al., 2015b). With increasing temperature, the primary products are broken down into smaller molecules through secondary cracking processes and lead to generation of gas and pyro-bitumen as end products (Mani et al., 2015a,b).

The amount Q of oil and gas formed is expressed as follows:

$$Q = \sum_{i=1}^{N} (Xio - Xi) \tag{3.3}$$

where Xio is the value of Xi at $t = 0$.

Kinetic parameters for thermal cracking of kerogen were obtained using the Optkin program from IFP/Beicip, Franlab. It utilizes the results of Rock Eval pyrolysis performed on the source rocks and calibrates the kinetic parameters of thermal cracking of organic matter into hydrocarbons (Espitalie et al., 1987; Ungerer and Pelet, 1987; Mani et al., 2015b). Using the Rock Eval pyrolysis time, temperature, and heating rates and HI as input data, the best fitting kinetic parameters (Arrhenius constants A, activation energies E, and initial petroleum potentials, Xio) for the observed S2 (hydrocarbon peak obtained during cracking of kerogen) are calculated (Espitalie et al., 1987; Mani et al., 2015b). The best adjustment corresponds to the minimum of the error function, which is defined by the summation of the quadratic differences calculated between measured and computed values (Espitalie et al., 1987; Mani et al., 2015b). The kinetic parameters are used to predict the amount of hydrocarbons generated as a function of temperature and time (Tissot and Espitalie, 1975; Mani et al., 2015b).

3.4 GEOCHEMICAL CHARACTERIZATION OF ORGANIC MATTER

Preliminary geochemical investigations and follow-up analyses using sophisticated techniques such as programmed pyrolysis, bulk and compound specific stable isotopic composition measurements, biomarker separation and analysis, etc., are used to evaluate the quality of organic matter in shale rocks and the quantitative generation of hydrocarbons.

3.4.1 Rock Eval Pyrolysis

Rock Eval pyrolysis is one of the most basic screening steps in evaluation of a source rock and is used to estimate the petroleum potential of source rocks. It involves an open system cracking of sedimentary organic matter according to a programmed temperature pattern (Mani et al., 2015b). The entire process takes place in two ovens, pyrolysis and oxidation (combustion), respectively of the Rock Eval pyrolyzer (Fig. 3.3). The pyrolyzed hydrocarbons, both the thermo-labile ones (forming the peak S1) and those obtained during the cracking of organic matter (forming the peak S2), are detected by a flame ionization detector (FID) (Behar et al., 2001; Mani et al., 2015a). The combustion of the residual rock, recovered after pyrolysis, takes place in oxidation oven under artificial air (N_2/O_2) up to 850°C. During pyrolysis and combustion, the released CO and CO_2 are monitored online by an infrared cell (Behar et al., 2001; Mani et al., 2015a). This enables the determination of organic and mineral carbon content of samples, defined as the TOC and MinC, respectively (Behar et al., 2001; Mani et al., 2015b). The T_{max} value is a thermal maturity parameter (Mani et al., 2015b). It corresponds to the temperature at which the maximum amount of hydrocarbons are released from the thermal degradation of kerogen, i.e., the temperature at which S2 peak reaches its maximum. Among the other calculated parameters of Rock Eval, the hydrocarbon

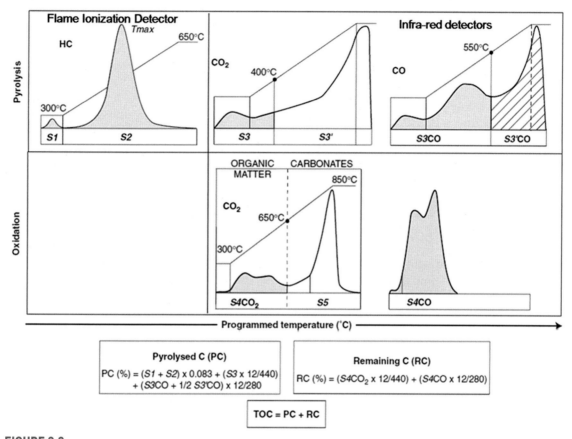

FIGURE 3.3
Rock Eval pyrolysis parameters for source rock evaluation (Behar et al., 2001).

potential or HI is defined by $100 \times S2/TOC$. The OI is defined as $100 \times S3/TOC$, where S3 is the CO_2 released during the pyrolysis. These indices help in establishing the kerogen type and its maturity (Behar et al., 2001; Espitalié et al., 1987; Mani et al., 2014, 2015b; Peters and Cassa, 1994).

3.4.2 Experimental Parameters for Rock Eval Pyrolysis

Pyrolysis parameters of shales have been obtained using the Rock Eval 6 pyrolyzer, Turbo version from Vinci Technologies (Mani et al., 2015a). The performance of the pyrolyzer is checked using the IFP standard, 160,000 ($S2 = 12.43$; $T_{max} = 416\,°C$). The samples were weighed in pre-oxidized crucibles, depending upon the organic matter content (~50–70 mg of shale; and 8–15 mg of coaly-shale). The shale samples were run in analysis mode using the bulk rock method and basic cycle of Rock Eval 6. The data is reported on a dry weight basis (Mani et al., 2014, 2015a).

3.4.3 Closed System Pyrolysis

A closed system, offline pyrolysis of shales was performed to desorb the thermo-labile gases present in the rock matrix (Fig. 3.4; Mani et al., 2015b). 500 mg of powdered shale was heated in the evacuated pyrolysis assembly at a fixed temperature of 300°C for 3 min (Fig. 3.4). Saturated KOH solution was added externally to a glass ampoule to absorb the carbon dioxide released during the pyrolysis of shales (Mani et al., 2015b). The thermally desorbed gases released during the pyrolysis were collected by upward displacement in the graduated limb of an apparatus fitted with a silicone septum. The desorbed gases were analyzed for their stable carbon isotope ratio compositions (Mani et al., 2015b).

3.4.3.1 Gas Chromatography

Gas Chromatography (GC) is used for the qualitative and quantitative identification of chemical compounds and essentially includes the separation and identification of chemical mixtures (Grob, 2004; Mani et al., 2015a). It consists of three main components: 1) an injector, which is a port meant for injecting the samples into the GC, 2) a column in which the analyte gets separated into individual components, depending upon its affinity with the stationary phase and the mobile carrier gas phase, and 3) the detector, where the composition and concentration of the analyte are determined (Mani et al., 2015a). The molecules of the analyte are partitioned between the stationary phase present in the column and the mobile

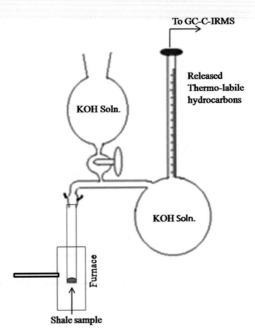

FIGURE 3.4

A schematic of the offline pyrolysis assembly used for collecting the thermo-labile gases from shales.

phase of the carrier gas. A carrier gas is an inert gas such as nitrogen or helium that carries the sample through the column to the detector. The column comprises the stationary phase, which is made up of polymeric material (Mani et al., 2015a). It is kept in a heated oven for maintaining the column temperature. Depending upon analyte composition, different types of columns, such as capillary or packed, with specific polymers are used. The components are eluted by the mobile phase. They reach the detector at different times, which is very specific for each component under a particular condition of pressure and temperature. It is called the retention time (Rt) of a respective component (Mani et al., 2015a). The hydrocarbon concentration is detected using the FID. Here, the organic compounds are burnt in a flame of hydrogen and air. A collecting electrode collects the electrons produced during combustion. The resulting current is the response of the detector in the form of series of signal peaks. The concentration is determined using the peak area or peak height as basis (Grob, 2004; Mani et al., 2015a). The GC, when hyphenated with a mass detector, becomes a more powerful technique for trace level organics in sediments than GC-FID.

3.4.4 Mass Spectrometry
3.4.4.1 *Gas Chromatography-Mass Spectrometry*

Gas chromatography–mass spectrometry (GC–MS) is used to detect compounds using the relative gas chromatographic retention times and elution patterns of components of a mixture in combination with the mass spectral fragmentation patterns, which is the characteristic of a compound's chemical structures (Sneddon et al., 2007; Mani et al., 2015a). A typical GC–MS system performs the following functions: 1) separation of individual compounds in a mixture by gas chromatography; 2) transfer of separated components to the ionizing chamber; 3) ionization; 4) mass analysis; 5) detection of the ions by an electron multiplier; and 6) data acquisition, processing, and display by a computer system (Mani et al., 2015a). As the individual organic compounds elute from the GC column, they enter the MS (Mani et al., 2015a). During ionization, they are bombarded by a stream of electrons leading to the fragmentation into ions. The mass of the fragment divided by its charge is the mass to charge ratio (M/Z). Almost always, the charge is +1, and M/Z ratio represents the molecular weight of the fragment (Mani et al., 2015a). In general, MS are configured with magnetic sector or quadrupole type mass analyzers. A quadrupole GC–MS has a group of four electromagnets that focus each fragment through a slit into the detector. These are programmed to direct only certain fragments, one at a time (scan), until the complete range of M/Z is recovered (Mani et al., 2015a). This produces a mass spectrum, which is a graph of the signal intensity (relative abundance) versus the M/Z ratios (essentially the molecular weight) (Mani et al., 2015a). Each compound has a unique fingerprint, and software is equipped with a library of spectra for unknown compounds (Sneddon et al., 2007; Mani et al., 2015a).

3.4.4.2 Extraction of Organic Matter

Extraction of organic matter from shales is carried out using the Buchi Speed Extractor E–914 (Mani et al., 2016). It performs the extraction using organic solvents at raised temperature and pressure. 3 g of sample was mixed with 3 g of sand and placed between the top and bottom layers of 1.5 g sand each, in a 40-mL extraction cell (Mani et al., 2016). The extraction was carried out in an atmosphere of nitrogen with a maximum temperature and pressure of 100°C and 100 psi, respectively, using dichloromethane (DCM) and methanol in ratio of 9:1 (Mani et al., 2016). The Speed Extractor was programmed for three extraction cycles. The collected organic extracts were reduced to about 1 ml on the Buchi Rotavap V-855. *n*-Pentane was added to the extract for the separation of asphaltenes. HCl-treated Cu turnings were added to remove the sulfides. The aliphatic and aromatic fractions were separated using silica gel column chromatography. The aliphatic fraction was eluted using 100 mL of petroleum ether (PE) and aromatic fraction using 150 ml of PE:DCM (1:4) mixture. The solvent fractions containing the analytes were reduced to 0.5–1 ml under the gentle stream of N_2 gas. Internal standards, 5(H) androstane, and d_{10}-naphthalene were added to the alkane and aromatic fraction, respectively, for the quantification of target analytes.

3.4.4.3 Operational Parameters for GC–MS

n-Alkanes were analyzed by a Clarus 500 GC–MS from Perkin Elmer, equipped with Rxi-5ms column (low polarity, fused silica) of length 60 m. The injector port was maintained at a temperature of 300°C. The GC oven was programmed at a temperature of 40°C (held for 1 min), and raised to 130°C at 8°C/min (held for 1 min), and finally to 300°C at 5°C/min and held for 25 min. Helium (99.999% purity) at a constant flow of 1.0 mL/min was used as a carrier gas. The mass spectrometer was operated in electron ionization (EI) I mode at 70 Ev. The temperature of the MS transfer line and ion source was set at 240 and 280°C, respectively. The filament emission current was 25 μA, and the filament-multiplier delay was set at 12 min. 1 μL of sample was injected into the GC–MS through the auto-sampler. The GC–MS was operated in full scan mode, covering a mass range (*m/z*) of 50–600 amu. The *n*-alkanes and isoprenoids were assigned from *m/z* 57 and 183 chromatograms, respectively. Compound identification was performed by comparing the GC retention times, fragmentation pattern, and *m/z* peaks of ions with those of commercially available standard from Chiron AS, Norway, and published mass spectra (Peters et al., 2005). The quantification was done by comparing the peak areas of target compounds with those of the internal standards for alkanes. For the PAH analysis, the GC–MS oven was programmed at an initial temperature of 60°C with hold of 1 min, and then ramped in three stages-initially to 175°C at a rate of 6°C/min with a hold of 4 min, followed by 235°C at 3°C/min with a hold of 4 min, and finally to 300°C at 5°C/min, with a hold of 15 min. GC–MS was operated in full scan mode, scanning from m/z 50–600 amu. The methyl, dimethyl, and trimethyl naphthalenes were assigned from the m/z 142, 156, and 170 chromatograms, respectively; and phenanthrene, dibenzothiophene, and methyl and dimethyl phenanthrenes from m/z 178, 184, 192, and 206 chromatograms, respectively.

3.4.4.4 Isotope Ratio Mass Spectrometry

Isotope ratio mass spectrometry (IRMS) has been widely used to determine the ratio of stable isotopes of several low molecular weight elements such as carbon ($^{13}C/^{12}C$) and oxygen ($^{18}O/^{16}O$) in geological samples (Mani et al., 2015a). Hyphenated techniques include the separation power of a gas chromatograph coupled to the mass spectrometer, along with the introduction of a sample combustion interface into the gas chromatograph-IRMS (Platzner, 1997 (Mani et al., 2015a). Broadly, a mass spectrometer comprises the 1) ion source for the fragmentation of the sample molecule into ions and 2) mass analyzer for separating and detecting the ion beam according to the mass of the respective ions (Mani et al., 2015a). In GC–C–IRMS, the separated products of the sample mixture carried in the stream of helium are eluted at the output of the gas chromatograph and pass through an oxidation/reduction reactor and then are introduced into the ion source of mass spectrometer for the ratio determination (Mani et al., 2015a). An open split-coupling device ensures that only a part of the sample/reference gas containing carrier gas is fed into the ion source, thus reducing the volume constraints and sample size (Mani et al., 2015a).

Continuous Flow-IRMS is a standard term used for the IRMS instruments that are coupled online to preparation lines or instruments (Mani et al., 2015a). This includes the 1) Gas Chromatography–Combustion–IRMS (GC–C–IRMS), used for the compound specific isotope ratio determination of individual hydrocarbon components, 2) Gas Bench–IRMS (GB–IRMS), used for the C and O isotope ratio determinations on carbonates, and 3) Elemental Analyzer–IRMS (EA–IRMS), for the determination of coexisting organic matter in rocks (Mani et al., 2015a).

3.4.4.5 GC–C–IRMS

An Agilent 6890 GC coupled to a Finnigan-Delta Plus^{XP} IRMS via a GC combustion III interface was used to study the carbon isotope ratios (Mani et al., 2015a). 1 mL of the desorbed gas was injected into the GC, which was equipped with a "Pora Plot Q" capillary column, 25 m in length and a diameter of 0.32 mm, in splitless mode with helium as the carrier gas at a fixed oven temperature of 28°C (Mani et al., 2015b). The hydrocarbon gases were separated chromatographically, and after elution from GC column, they entered a pre-oxidized Cu-Ni-Pt combustion reactor maintained at 960°C, where they were converted into carbon dioxide and water. The purified CO_2 after combustion entered into the mass spectrometer for $^{13}C/^{12}C$ ratio measurement of the respective hydrocarbon component. The GC–C–IRMS was calibrated to the international standard Vienna Pee Dee Belemnite (VPDB) using Natural Gas Standard (NGS-1) mixture (Mani et al., 2011, 2015a, 2015b).

3.4.4.6 Bulk Carbon Isotope Analysis

Bulk carbon isotope analysis ($\delta^{13}C_{org}$) of the HF-HCl extracted kerogen concentrate was performed on a Thermo Finnigan Flash-Elemental Analyzer (TC/EA) connected to Delta [plus]XP isotope ratio mass spectrometry (IRMS) (Mani et al., 2015b). The samples were held in tin containers and placed inside an autosampler drum, where they were purged with a continuous flow of helium gas and then dropped into a vertical quartz tube maintained at 1020°C at preset intervals (Mani et al., 2015b). Flash combustion of the sample took place in a helium stream temporarily enriched with pure oxygen. The resulting gas mixture entered the chromatographic column (Porapak PQS) where the individual components were separated. The isotopic composition of evolved carbon dioxide was measured using IRMS (Mani et al., 2015b). The calibration of Flash EA–IRMS was done using international standards, NBS-20, and graphite. The precision was found to be within ±0.3‰ of the reported value (Mani et al., 2015b).

3.5 DISCUSSION

The organic matter in gas shales is subjected to various physical, chemical, and biological processes, which lead to generation of natural gas. Various primary and secondary thermal decomposition reactions act on the organic matter along with the biogenic degradation processes, leading to the generation of gas in gas shales. It is stored in the form of adsorbed gas in organic matter and clays, as free gas held in the pores or micro-pores, and in the spaces and crevices created by the natural rock cracking such as fractures or micro-fractures. The solution gas is dissolved within bitumen and oil. The amount of adsorbed methane is a function of organic matter as well as the surface area of organic matter and/or clays. It usually increases with increase in these parameters. If the free gas content is high in shale gas wells, generally it results in higher initial rates of production. The free gas resides in fractures and pores, so it becomes easier to move it out relative to adsorbed gas. As adsorbed gas is slowly released from the gas shales, the initial flow rates decline rapidly to a low, steady rate.

When compared to the conventional reservoirs like sandstone, limestone, or dolostone, the gas shales have very low permeability. The effective bulk permeability is typically much less than 0.1 millidarcy (md) in gas shale. However, exceptions do exist where there are naturally fractured shale rocks like that of the Antrim shale, Michigan Basin, USA. In most cases, the well is artificially stimulated or fractured to increase permeability. This helps to produce gas in economical quantities.

A closer look at the shale gas system spurs into consideration as to where and to what extent are the organic-rich rocks present, the quantity of organic matter it contains, type of organic matter (gas vs. oil-rich shale), clay and other minerals it contains, depth of burial and cooking, its brittleness versus ductility (break or bend), and its fracture system (natural fractures). All of these characteristics change in a shale formation across its areal extent. Increasing research on shale gas has led to the establishment of certain characteristics of potential and prolific plays (Mani et al., 2015b). In general, the productive gas shales are >200 ft (65 m) thick, have TOC content >3 wt% and HI > 350 mg HC/gTOC, and contain Type II marine kerogen of thermal maturity >1.10% Ro (Curtis, 2002; Jarvie et al., 2007; Loucks and Ruppel, 2007; Horsfield and Schulz, 2012; Mani et al., 2015b and several others).

3.5.1 Case Studies
3.5.1.1 Damodar Basin, India

The organic-rich and thermally mature Gondwanan shales of the Indian subcontinent are a likely source for the gaseous hydrocarbons (Mani et al., 2015b). Gas shows have been encountered in the wells drilled in the Raniganj area of Damodar Valley basin, which forms an important coal repository among the Gondwana basins of India (Padhy and Das, 2013; Mani et al., 2015b). Barakar and Raniganj are the main coal-bearing formations in the basin, and a marine/lacustrine succession that got deposited between these continental depositions resulted in the coal-devoid Barren Measure formation (Chandra, 1992; Mani et al., 2015b). Thermal maturity of the coals surrounding the Barren Measure shale formation suggests it to be within the gas window (Padhy and Das, 2013), making it a potential shale gas target (Mani et al., 2015b). The organic matter in shales from the Jharia sub-basin of Damodar Valley show excellent organic richness. The TOC content ranges between 2.86% and 23.09%. Organic matter is characterized by Type II/III and Type III kerogen, and the thermal maturities span between the mature (oil window-gas condensate) and post-mature (dry gas) zone for hydrocarbon generation (Mani et al., 2015b). The Raniganj and Barren Measure shales are in the late oil generation window, and the Barakar shales show post-mature stage for oil (Fig. 3.5, Mani et al., 2015a,b). The bulk and compound specific carbon isotope ratios of hydrocarbon gases from the flash combustion and thermal desorption studies, respectively, of these shales indicate a thermogenic gas from the terrestrial organic matter, except for Barren Measure, where a lacustrine/marine source has been inferred (Mani et al., 2015b). The isotopic compositions vary in a narrow range and are nearly similar for most of the studied samples (−40.4 to −44.8‰). The exposure of organic matter to high subsurface temperatures has not enriched or altered the $\delta^{13}C$ values significantly (Mani et al., 2015b).

The shale organic extracts contain high proportions of saturated hydrocarbons in the range nC_{13} to nC_{30} and acyclic isoprenoids in all the samples (Fig. 3.6). The aromatic fractions in the organic extracts of the Raniganj, Barren Measure, and Barakar shale, in general, contain naphthalene, phenanthrene, biphenyl, and their alkylated derivatives, along with a low abundance of dibenzothiophene and its homologs (Fig. 3.7). The ratio obtained by dividing the sum of odd carbon-numbered alkanes to the sum of even carbon-numbered alkanes has been defined by CPI (Didyk et al., 1978). It gives information on the maturity of the organic matter. The CPI of Damodar basin shales range from 0.9

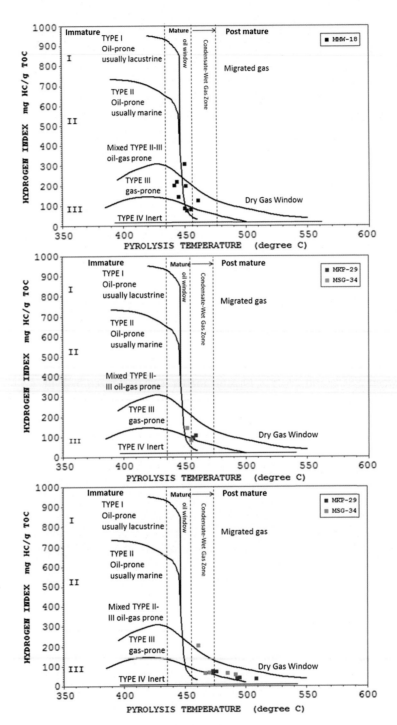

FIGURE 3.5

HI versus Tmax plots (after Espitalie et al., 1987) indicating the kerogen type and thermal maturity of shales from the Jharia sub-basin: (A) Raniganj, (B) Barren Measure, and (C) Barakar Formations (Mani et al., 2015b).

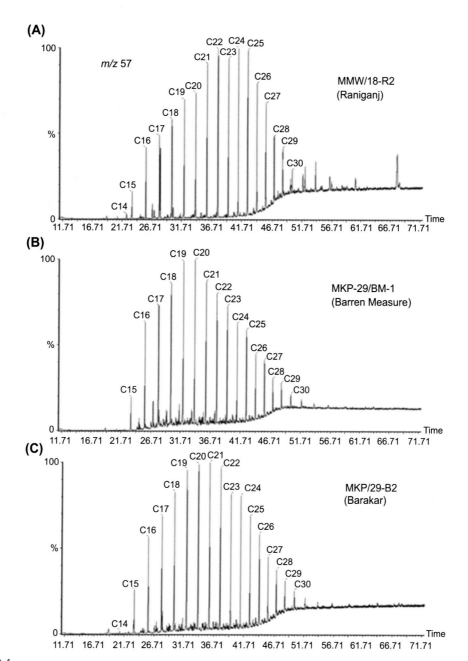

FIGURE 3.6
Representative n-alkane spectra of the extracted organic matter from the shales of (A) Raniganj, (B) Barren Measure, and (C) Barakar Formations of Damodar Valley basin.

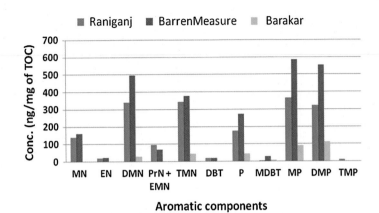

FIGURE 3.7

Concentration distribution of aromatic biomarkers in the extracted organic matter of the shales from the Damodar Valley basin.

to 1.1, which indicates no even or odd carbon preference and suggests the organic-rich sediments to be thermally mature. This is also corroborated by the unimodal peak distribution of the *n*-alkanes. The differences of Pr/Ph ratios of the Damodar Valley shales therefore indicate a fluvio-marine and coastal swamp depositional environment under oxic to sub-oxic conditions. Because the ratios of α/β isomers are maturity dependent, the concentration of aromatic isomers was applied in the calculation of temperature-sensitive maturity parameters such as methylphenanthrene indices (MPI-1), dimethylnaphthalene ratios (DNR-1 and DNR-2), and trimethylnaphthalene ratio (TNR-1). The ratios of the Damodar basin shales in general are high, suggesting a mature to highly mature state of kerogen (Fig. 3.8). This is also supported by the Tmax parameter of Rock Eval pyrolysis (Fig. 3.9).

Vitrinite and inertinite are the dominant macerals in the Jharia shales. Liptinite occurs sparsely (Mani et al., 2015b). The Raniganj and Barren Measure shales show vitrinite and inertinite macerals, with lesser occurrence of liptinite. The random reflectance (Rr%) varies between 0.99 and 1.22 for Raniganj shales and 1.1–1.41 for the Barren Measure. The Barakar shales consist mainly of vitrinite and inertinite with a random reflectance ranging between 1.11 and 2.0. The mineral matrix is characterized by abundant clay minerals with siderite and pyrite (Fig. 3.10; Mani et al., 2015b).

The kinetic parameters obtained during pyrolysis studies indicate that Raniganj has lower activation energies (ΔE = 42–62 kcal/mol) in comparison to Barren Measure and Barakar (ΔE = 44–68 kcal/mol) (Fig. 3.11, Mani et al., 2015b). Temperature for onset (10%), middle (50%), and end (90%) of

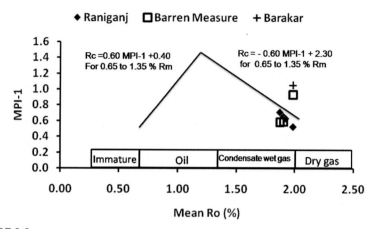

FIGURE 3.8

Methyl phenanthrene index (MPI-1) vs. calculated vitrinite reflectance indicating the maturity of shales from the Damodar Valley basin.

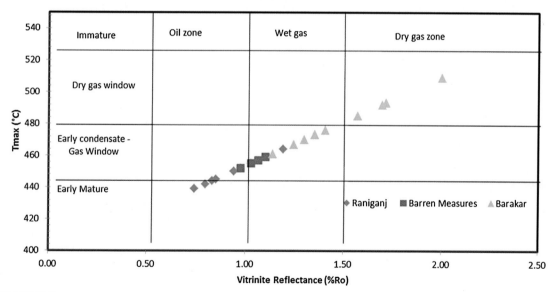

FIGURE 3.9

Correlation plot of vitrinite reflectance and Rock Eval Tmax data from the shales of Jharia sub-basin, Damodar Valley (Mani et al., 2015b).

kerogen transformation is least for the Raniganj shales (Mani et al., 2015b). It is followed by the Barren Measure and Barakar shales The Permian Gondwana shales, in particular, those from the Barren Measure formation, demonstrate excellent properties of a potential shale gas system (Mani et al., 2015b).

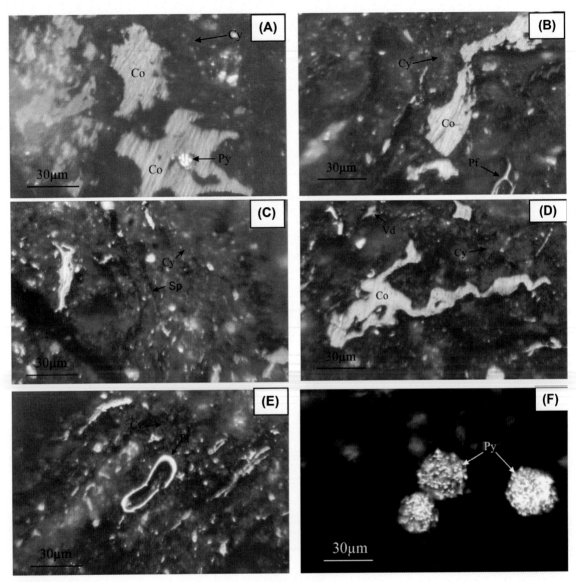

FIGURE 3.10

Photomicrographs of Barren Measure shale (MKP-29/BM-1) showing the macerals - collotelinite (Co), pyrite (py) - (A); collotelinite (Co), clay (Cy), pyrofusinite (Pf) - (B); clay (Cy), sporinite (Sp)- (C); vitrodetrinite (Vd), collotelinite (Co), clay (Cy), - (D); clay (Cy), pyrofusinite (Pf)- (E); pyrite (Py) - (F) under reflected light (magnification of 625×) (Mani et al., 2015b).

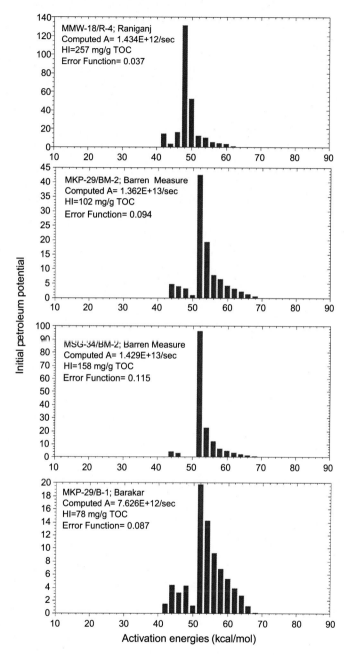

FIGURE 3.11

The frequency factors and activation energy distributions for the thermal cracking of organic matter in shales from the Jharia sub-basin, Damodar Valley (Mani et al., 2015b).

3.5.1.2 Pranhita–Godavari Basin

The Pranhita-Godavari (P–G) basin is a large, inter-cratonic, Gondwana basin located in the eastern part of Indian subcontinent. The nearly 3 km sedimentary thickness of the basin ranges in age from the Late Carboniferous/Early Permian to Cretaceous. The stratigraphic succession of the Gondwana sediments comprise Talchir, Barakar, Barren Measures belonging to the Lower Gondwana and Kamthi, Maleri, Kota, Gangapur, and Chikiala Formations belonging to the Upper Gondwana. Lithologically, Barakar, Barren Measure, and Kamthi Formations consist of gray shales, dark gray shales, and carbonaceous shales with thickness varying from 400 to 800 m. The preliminary geochemical studies of shales of Pranhita–Godavari basin indicate that Barakar shales contain abundant humic organic matter deposited in an intermediate to swampy environment, associated with immature Type III kerogen favorable for gas generation (Pal et al., 1992). With the discovery of gas shows in Krishna–Godavari basin, where the source

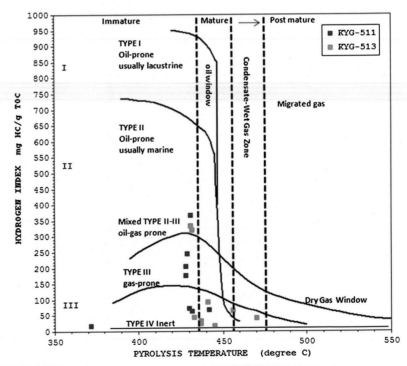

FIGURE 3.12

Hydrogen index (HI) versus Tmax for shale samples of Barakar Formation from KYG-511 and KYG-513 boreholes.

rocks are reported to be the Permo–Carboniferous sediments of Pranhita–Godavari basin (Sudhakar, 1990), the organic geochemical characterization of shales in P–G basin are significant to assess the shale gas prospectivity of the basin.

Based on organic geochemical analysis of drill cores of shales belonging to Kamthi and Barakar formations from three exploratory boreholes of Pranhita–Godavari coalfields, it is inferred that both the formations possess high TOC up to 35% with an average of 7.4%. These suggest excellent source rock potential of shales. HI versus Tmax plot of Barakar and Kamthi shales reveal that the samples are dominated by Type II-III oil-gas-prone kerogen and Type III gas-prone kerogen (Figs. 3.12 and 3.13, repectively). The thermal maturity of kerogen ranges from the immature to mature stage of hydrocarbon generation. The shales from Barakar and Kamthi formations of P–G basin exhibit substantial potential for generating oil and thermogenic gas upon thermal cracking (Figs. 3.12 and 3.13).

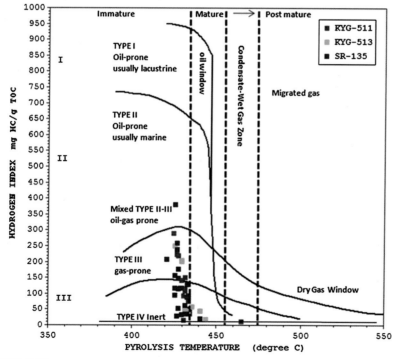

FIGURE 3.13

Hydrogen index (HI) versus Tmax for shale samples of Kamthi Formation from KYG-511, KYG-513, and SR-135 boreholes.

3.6 SUMMARY

The conventional source rocks have been studied for their organic content over the last several decades. The recent move to explore and exploit the shale gas and shale oil has necessitated a resurgence of geochemical methods for the appraisal of shale rocks. Attributes like organic richness, type of kerogen, and its thermal maturity are directly associated with the hydrocarbon generation potential of a source rock. The expanded utilization of established techniques and the development of newer ones are taking the exploratory activities forward in terms of precise delineation of source rocks kitchens.

References

Alexander, R., Kagi, R.I., Rowland, S.J., Sheppard, P.N., Chirila, T.V., 1985. The effects of thermal maturity on distribution of dimethylnaphthalenes and trimethylnaphthalenes in some ancient sediments and petroleums. Geochim. Cosmochim. Acta 49, 385–395.

Behar, F., Beaumont, V., Penteado, H.L., De, B., 2001. Rock-Eval 6 technology: performances and developments oil & gas science and technology. Rev. IFP 56 (2), 111–134.

Bordenave, M.L., 1993. Applied Petroleum Geochemistry. Technip.

Braun, R.L., Burnham, A.K., 1987. Analysis of chemical reaction kinetics using a distribution of activation energies and simpler models. Energy Fuels 1, 153–161.

Curtis, J.B., 2002. Fractured shale-gas systems. AAPG Bull. 86 (11), 1921–1938.

Chandra, D., 1992. Jharia Coalfield. Mineral Resources of India 5. Geological Society of India, Bangalore, p. 149.

Didyk, B.M., Simoniet, B.R.T., Brassell, S.C., Eglington, G., 1978. Organic geochemical indicators of paleoenvironmental conditions of sedimentation. Nature 272, 216–222.

Espitalie, J., Marquis, F., Sage, L., 1987. Organic geochemistry of the Paris Basin. In: Brooks, J., Glennie, K. (Eds.), Petroleum Geology of North West Europe. Graham and Totman, London, pp. 71–86.

Grob, R.L., 2004. Theory of chromatography. In: Grob, R.L., Barry, E.F. (Eds.), Modern Practice of Gas Chromatography, fourth ed. John Willy & Sons. Co, New York. 743p.

Horsfield, B., Schulz, H.M., 2012. Shale gas exploration and exploitation. Mar. Petroleum Geol. 31 (1), 1–2.

Hunt, J., 1996. Petroleum Geology and Geochemistry.

International Committee for Coal and Organic Petrology (ICCP), 2001. The new inertinite classification (ICCP System 1994). Fuel 80, 459–471.

Jarvie, D.M., Hill, R.J., Ruble, T.E., Pollastro, R.M., 2007. Unconventional shale gas systems: the Mississippian Barnett Shale of north-central Texas as one model for thermogenic shale-gas assessment. AAPG Bull. 91, 475–499.

Jones, R.W., 1983. Organic Matter Characteristics Near the Shelf-slope Boundary, vol. 33. Society of Economic Paleontologists and Mineralogists(SEPM), Special Publication, pp. 391–405.

Jones, R.W., Demaison, G.J., 1982. Organic facies-stratigraphic concept and exploration tool. In: Salvidar-Sali, A. (Ed.), Proceedings of the Second ASCOPE Conference and Exhibition, October 1981, Manila, pp. 51–68.

Lijmbach, G., 1975. On the origin of petroleum. In: Proceedings of the 9th World Petroleum Congress, vol. 2. Applied Science Publishers, London, pp. 357–369.

Loucks, R.G., Ruppel, S.C., 2007. Mississippian Barnett shale: Lithofacies and depositional setting of a deep-water shale-gas succession in the Fort Worth basin, Texas. AAPG Bull. 91 (4), 579–601.

Mani, D., Patil, D.J., Dayal, A.M., 2011. Stable carbon isotope geochemistry of near surface adsorbed alkane gases in Saurashtra basin. Chem. Geol. 280 (1–2), 144–153.

Mani, D., Dayal, A.M., Patil, D.J., Hafiz, M., Hakoo, N., Bhat, G.M., 2014. Gas potential of proterozoic and phanerozoic shales from NW Himalaya, India: inferences from pyrolysis. Int. J. Coal Geol. 128-129, 81–95.

Mani, D., Patil, D.J., Dayal, A.M., 2015a. Organic properties and hydrocarbon generation potential of shales from few sedimentary basins of India, petroleum geosciences: Indian context. Springer Ser. Geol. 99–126.

Mani, D., Patil, D.J., Dayal, A.M., Prasad, B.N., 2015b. Thermal maturity, source-rock potential and kinetics of hydrocarbon generation in Permian shales of Damodar Valley Basin, Eastern India. Mar. Petroleum Geol. 66, 1056–1072.

Mani, D., Ratnam, B., Kalpana, M.S., Patil, D.J., Dayal, A.M., 2016. Elemental and organic geochemistry of Gondwana sediments from the Krishna–Godavari basin, India. Chem. Erde Geochem. 76 (1), 117–213.

Mishra, S., Mani, D., Kavitha, S., Patil, D.J., Kalpana, M.S., Vyas, D.U., et al., 2015. Pyrolysis results of shales from the south cambay basin, India: implications for gas generation potential. J. Geol. Soc. India 85, 647–656 June 2015.

Moldowan, J.M., Seifert, W.K., Gallegos, E.J., 1985. Relationship between petroleum composition and depositional environment of petroleum source rocks. Am. Assoc. Pet. Geol. Bull. 69, 1255–1268.

Padhy, P.K., Das, S.K., 2013. Shale oil and gas plays: Indian sedimentary basins. Geohorizons 18 (1), 20–25.

Pal, M., Venkatesh, V., Balyan, A.K., Sarkar, A., 1992. Hydrocarbon prospects of Gondwana basins in India: source rock studies of Kamthi sub-basin of Pranhita–Godavari Graben. J. Geol. Soc. India 40, 207–215.

Peters, K.E., Cassa, M.R., 1994. Applied source rock geochemistry. In: Magoon, L.B., Dow, W.G. (Eds.), The Petroleum System—From Source to Trap: Tulsa, Okla., American Association of Petroleum Geologists Memoir 60, pp. 93–117.

Peters, K.E., Walters, C.C., Moldowan, J.M., 2005. The Biomarker Guide. Biomarkers and Isotopes in the Environment and Human History, vol. 1. Cambridge Press.

Petsch, S.T., 2014. Weathering of organic carbon. In: Reference Module in Earth Systems and Environmental Sciences, from Treatise on Geochemistry, vol. 12, second ed., pp. 217–238.

Platzner, I.T., 1997. Modern Isotope Ratio Mass Spectrometry. Wiley, Chichester, U.K.

Romero, A.M., Philip, R.P., 2012. Organic geochemistry of the Woodford Shale, southeastern Oklahoma: How variable can shales be? AAPG Bull. 96 (3), 493–517.

Schoell, M., 1983. Genetic characterization of natural gases. AAPG Bull. 67, 2225–2238.

Seifert, W.K., Moldowan, J.M., 1986. Use of biological markers in petroleum exploration. In: Johns, R.B. (Ed.), Chapter 7, Methods in Geochemistry and Geophysics. Elsevier, Amsterdam, pp. 261–290.

Sneddon, J., Masuram, S., Richert, J.C., 2007. Gas chromatography-mass spectrometry-basic principles, instrumentation and selected applications for detection of organic compounds. Anal. Lett. 40 (6), 1003–1012. http://dx.doi.org/10.1080/00032710701300648.

Sudhakar, R., 1990. Scope of integrated exploration research in Krishna–Godavari, Cauvery and Andaman basins. In: Pandey, J., Banerjie, V. (Eds.), Proceedings of the Conference on 'Integrated Exploration Research, Achievements and Perspectives', pp. 69–80.

Sýkorová, I., Pickel, W., Christianis, K., Wolf, M., Taylor, G.H., Flores, D., 2005. Classification of huminite – ICCP system 1994. Int. J. Coal Geol. 62, 85–106.

Tissot, B.P., Welte, D.H., 1984. Petroleum Formation and Occurrence. Springer-Verlag, New York.

Tissot, B.P., Espitalie, J., 1975. Thermal evolution of organic matter in sediments: application of a mathematical simulation. Petroleum potential of sedimentary basins and reconstructing the thermal history of sediments. Rev. Inst. Fr. Petr. 30 (5), 743e778.

Tissot, B., Pelet, R., Ungerer, P., 1987. Thermal history of sedimentary basins, maturation indices, and kinetics of oil and gas generation. 71 (12), 1445–1466.

Ungerer, P., Espitalié, J., Marquis, F., Durand, B., 1986. Use of kinetic models of organic matter evolution for the reconstruction of paleo-temperatures. In: Burrus, J. (Ed.), Application to the Case of the Gironville Well (France). Thermal Modeling in Sedimentary Basins, Technip, Paris, pp. 531–546.

Ungerer, P., Pelet, R., 1987. Extrapolation of the kinetics of oil and gas formation from laboratory experiments to sedimentary basins. Nature 327, 52e54.

Van Krevelen, D.W., 1961. Coal: Typology-Chemistry-Physics-Constitution. Elsevier Science, Amsterdam.

CHAPTER 4

Basin Structure, Tectonics, and Stratigraphy

A.M. Dayal

CSIR-NGRI, Hyderabad, India

CONTENTS

4.1 INTRODUCTION

For sedimentary basins, a basic requirement is the area available for the accumulation of sediment. The sedimentary rocks are deposited by weathering of preexisting rock with various weathering agencies in different grain sizes and their transportation to the place of deposition. The preexisting rock could be basic or ultrabasic as well as acidic. The acidic rocks like granite will be weathered much faster than the basaltic rocks. The Earth consists of nine major plates and a few micro-plates. These plates are 100 km thick and are also known as lithospheric plates. The movement of these tectonic plates is responsible for the rifting, mountain building, subduction, volcanic activity, sea floor spreading, etc. The motion of these plates in geological history is responsible for the formation of major sedimentary basins all over the world. The basin formation is not only on land but also active in the ocean. Plate thickness of the crust varies from 30 to 75 km thick, but in the case of ocean sediment the crustal thickness is around 8–10 km.

55

Shale Gas. http://dx.doi.org/10.1016/B978-0-12-809573-7.00004-4

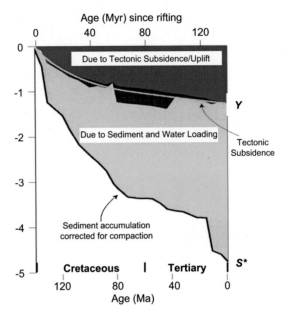

FIGURE 4.1
Subsidence of sediment in a marine system.

The tectonic activity forms the uplift and subsidence. In the case of subsidence the weathered material transported through wind or water accumulates. These areas of accumulation could be in lake beds, swamps, deltas, or estuary and shelf regions. A major amount of sediments are deposited in the marine region as the two-thirds of our Earth is marine and only one-third is terrestrial. In the subsidence region, once the sediment thickness increases due to sediment load the pressure of the sediment and temperature plays an important role in the formation of sedimentary rocks. Density of the accumulated sediments plays an important role in the formation of sedimentary rock (Fig. 4.1), and there is a relationship between the water depth and thickness of sediment it can accumulate. Total thickness of the sediment in the area could be 2.5 times the water depth.

Transport elements like wind, water in rivers, and the time frame are very important for the sediment accumulation. During flash floods and heavy rains, there is movement of very large rock particles, and such particles can be observed all along the rivers as rounded pebbles and boulders. The river is one of the most important transport media for weathering particles, and based on river velocity, various sizes of grains are deposited all along the river banks and also in low-lying areas during flash floods. The transportation of sediment by rivers forms the different sedimentary rock formations like breccia, conglomerate, sandstone, siltstone, shale, and clay rocks.

The weathered sediments accumulate either in shallow water or deep water depending on the grain size and environment of deposition. There are various types of sedimentary basins that are related to tectonic activity in the region. The concave-shaped basin is related with initially fast subsidence and later slow deposition. But if the initial subsidence is slow and later fast due to some tectonic activity, these are known as foreland basins or strike-slip basins related with rift. The subsurface heat in the upper mantle plays an important part in the rifting and subsidence of crustal rocks. The crustal thinning is responsible for rapid subsidence, and mantle plume activity is responsible for a slow rate of subsidence.

In a sedimentary basin, large amounts of sediment are collected with very good thickness, and these sediments are preserved for a long geological interval. The sediment deposited in any basin records the geological history of the region and also various tectonic activity and behavior of the environment. In the case of thick sedimentary deposits, pressure and temperature and presence of water plays an important role for biogenic activity. Shape of any sedimentary basin depends on the evolutionary history of the basin, which is related with plate motion, subduction, and rift activity. Organism activity also plays an important role for the sedimentary basin formation. Large amounts of limestone in different basins are a result of precipitation under biogenic activity.

Sedimentary basins can have various shapes and sizes. Tectonics in the sedimentary basins are related to rocks, climate and environment of deposition, and activity of microorganisms. Tectonics of the basin helps in accumulation of sediments. Sedimentary basins can be classified based on tectonics. On continental crust, basins are formed by divergent plate motion with extension and thermal effects. Such basins sag due to crustal thinning, thermal conditions, and excess sediments. The narrow elongate basins bounded by a large number of faults are responsible for basin formation in rift-related zones. Rift basins are related with up-doming of the lithosphere before the rifting. Thermal contraction due to cooling and deposition of a large amount of sedimentary material allows the formation of large rift basins. Rift basins are also related with the movement of two blocks apart, which is a common feature on the ocean floor. In the passive margin basins, two divergent continents separate, and oceanic crust forms in the rifted space. Oceanic sag basins form between a mid-ocean ridge by accumulating deep sea fan or basin sediments. Basins also form during the subduction process related with orogenic activity. Such basins are related to island arcs or an active continental margin. The oceanic basins are between microcontinents and part of the continental crust. The remnant basins consist of a large amount of sediments from rising areas and undergo synsedimentary deformation. Foreland basins are formed by depressing continental crust under the overload from orogenic activity. Strike-slip and wrench basins are associated with a tensional or compressional component. During rapid tectonic movements, a large amount of sediment is responsible for forming pre-depositional basins. They are filled by posttectonic activity.

4.2 PLATE MOTION

Plate motion is an important phenomenon of our Earth system. The various geological and geophysical activities on land and ocean are the result of plate motion. Plate boundaries can be divided as convergent, divergent, or transform. They are related with compression, extension, and strike-slip faults. Compression of plates is responsible for subduction activity. Mountain building is the result of subduction of one plate under another plate. There is a collision of continental plates and also continental plate to oceanic plate. In the case of compression of plates, there is crustal shortening, and the thickness of the crust increases. In the case of divergent activity of plates, new ocean floors are created and also related with large-scale volcanic activity and formation of new oceans. Convergence of oceanic plates with continental plates also results in mountain building activity. Motion of these plates is also responsible for major seismic activity as it activates many big fault systems. In fact, the active seismic zones at present are related with either tectonic activity at continental plates or the motion of a plate under continental or oceanic plates. In the case of divergent plate motion, new, smaller plates are moving apart. In the case of continental rifts, there is crustal thinning and faulting as the crust undergoes extension. Divergent movement of oceanic plates is responsible for the oceanic crustal thinning and formation of mid-ocean ridge basalt. The movement of plates is responsible for the formation of new continents and oceans in geological history.

4.3 RIFT BASINS

Rift basins are the result of crustal extension and formation of new oceanic lithosphere. Continent crustal extension is responsible for the formation of rift basins. There are many lakes formed due to continental rift. Activity of the mantle and production of a high heat flow zone under the crust develop active volcano systems. It has been observed that rifts form between two divergent plate boundaries with formation of new oceanic crust. Hot spots formed due to mantle plume activity also produce rifts on continental crust. The length of a rift basin depends on the thickness of lithosphere plates. There are cases of development of triple junctions and formation of new oceans. The shape and depth of a rift basin depends on the tectonic activity, type of rock, and geometry of the basin.

4.4 RIFT DEVELOPMENT

Geological history has a good record of rift basins. Rift basins occur in continental as well as in oceanic crust. A rift is related to a fault system, and in cross-section, a rift will be perpendicular to the fault system. Most of the rift basins, also known as grabens, are either filled with water in the form of lake or sediment in the form of rift basin. It has been observed that crustal stretching is always associated with the formation of rift valleys. A rift is the small structural disturbance in the crustal rocks that activates the associated fault system. Activation of a fault with another fault system activates the larger fault system and leads to crustal extension and thinning of Moho. This tectonic activity may be responsible for the uplift and formation of rift basins and new drainage systems. During crustal extension, mantle lithosphere becomes thinned, and the asthenosphere rises. As the rift cools down, there is post-rift subsidence, and the amount of subsidence is related to crustal thickness and the density of sediment. Thinning of continental lithosphere activates the upwelling of asthenosphere and generates melts, which are responsible for the uplift of crustal material. In the case of large-scale upwelling of asthenosphere, there will be large-scale volcanic activity.

In rift basins, initially, there will be subsidence due to sediment load, and the rate of subsidence depends on the thickness and density of sediments. This can also be seen in the Bouguer gravity anomaly of the basin. In the case of mountain regions, low gravity dominates, but high gravity is also associated, which provides information about subsurface sediments like ophiolites. There are many structural components active in rift basins. Faulted margins, uplifted flanks, hinged margins, troughs, and various faults will be part of the structural components, and some of them will be always active.

4.5 GEOLOGICAL ENVIRONMENTS ASSOCIATED WITH TECTONICS

In any geological environment, tectonics plays an important role for the drainage system, activation of minor faults, weathering patterns, and sedimentation. In the case of a rift basin, tectonics will be more active as these are the basins formed due to crustal extension. Divergent plates are a zone of active extension as the newly formed crust at the mid-ocean ridge system. Continent-to-continent collision spreads laterally, and after the collision, the thickness of crust increases, and gravity plays an important role in creating large-scale extension fault systems.

The geometry of the basin is related to the tectonic process active during that time. But the morphology is defined by the interplay of tectonic movements and sediment deposition. The basic principle of basin filling is the environment of deposition, supply of sediments, and geographic location of the basin. Rivers and wind are the media for transporting sediments to the basin. The alluvial deposits from sedimentary structures show flow patterns. The vertical sequence will be irregular as the environment of deposition changes with time. Floodplain deposits are suspended loads of silt and mud deposited in graded manner. Fluvial sediments are important as they are a source of hydrocarbons and ground water.

Lake sediments are related to waves and currents of water and are very sensitive to the environment. Deposition of limestone, iron stone, oil shales, and evaporites like gypsum are the product of a lacustrine environment. A fresh water lake has low total dissolved solids compared to a saline lake. Lakes are a good source of marine deposits. Lakes form with crustal subsidence in a rift zone or sagging of the continent due to tectonic forces. Lakes receive fine to very fine material that is transported into a deeper part of lake. Production of organic material in lakes is high due to enough nutrients in the form of bacteria and algae. Being a closed system, the deposition rate is high in lakes compared to the open marine environment. Lake sediments are a very good indicator for the changes in climate, as the climate is responsible for the deposition of sediment in lakes. Carbonaceous shales are a good source of hydrocarbon generation, and fine sediments deposited later act as a seal.

Stratigraphy of the basin is important to understand the different times of deposition and is also related to hydrocarbon deposition. Sediments deposited in shore regions are affected by sea level changes. The transition zone between the shore line and shallow sea plays an important role in sequence stratigraphy. At the estuary the river adjusts its gradient based on sea level. Lacustrine or marine deposits are related with the level of water in a lake or ocean. The water gradient in lakes away from a sea or ocean is controlled by local agents like wind and temperature. The periodicity of deposition has been observed. Accumulation of a large amount of sediments in a sedimentary basin is responsible for the subsidence of crust. When the lithosphere is heated by the lower asthenosphere, the mass does not change, but volume changes and results in uplift of continental crust. In the ocean, such activity is responsible for the oceanic ridges and oceanic spreading centers.

Inversion tectonics plays an important role for various stress patterns and forces acting on the plates. There is basin uplift due to increasing lithosphere thickness. A gravity anomaly in such a sedimentary basin has been observed. Negative gravity anomalies observed in the basin indicate thinning of crust due to extension. Compression of sediment is related with basin inversion, and the amount of compression will be related to the basin evolution and its extension. Inversion of a basin is related to relative uplift of the basin. Activation of faults as reverse and thrust faults are also related to inversion.

4.6 SEQUENCE STRATIGRAPHY

In any sedimentary cross-section of any basin, there will be alternating layers of different sedimentary rocks like limestone, shale, sandstone, unconformities, and again the same sequence. The sequence stratigraphy suggests that the deposition in the basin was not a continuous process of deposition, but it was related with the depositional environment, which includes terrestrial as well as marine. The weathering process, rainfall, drought, and tectonic events will be well recorded in the stratigraphy sequence. The thickness of sedimentary beds, mineral composition in the sedimentary beds, and the grain size will suggest an environment of deposition. Variation in sedimentary beds suggests abrupt changes in the deposition environment. Changes of grain size in the same sediment bed suggest slow changes in deposition, and in sediment cross-section, we have seen the same sediment sequence of the beds, but their thickness may vary and some of the bed may even be missing. These sedimentary sequences also record the sea level from the geological past.

Unconformity in the sedimentary sequence suggests structural disturbances either in the source region or in the basin. Change of sea water temperature, large-scale glaciation, or deglaciation will be very clear in the sediment sequence stratigraphy. In the geological past, sea level has changed a number of times, and these changes are well recorded in the sedimentary sequence. Even the large-scale basaltic activity due to migration of plates will be well recorded in these stratigraphy sequences.

In the stratigraphy sequence, there is a cycle of 20,000 and 40,000 years for change of sea level. Similarly, there is a cycle of glaciation in geological history. Some of these changes are related to the change of Earth's orbit from circular to elliptical. In sequence stratigraphy, very large cycles to small cycles of environment changes have been observed. The sequence stratigraphy has importance for mineral exploration, and study of this sequence stratigraphy has been carried out for a large amount of exploration work. For mineral exploration, hydrothermal activity plays an important role, and for hydrocarbon exploration, sequence of rock deposition and environment of sedimentation in the basin has importance.

4.7 FORMATION OF BASIN AND ECONOMIC IMPORTANCE

The various sedimentary basins are a good reservoir for minerals and hydrocarbons. Basin evolution and the environment of deposition play an important role in the type of mineral deposition. A basin is a subsidence that accumulates sediments through various transporting agencies. In a petroliferous basin the important aspects are history of basin and cycles of sediment accumulation. Various exploration techniques like geochemical, shallow seismic, and other geophysical parameters provide important clues about the presence of hydrocarbons in the basin. Geochemistry of the subsoil provides information about the presence of light hydrocarbons like methane, ethane, butane, and propane. The carbon isotopic study of these soil samples also provides source rock information. Shallow seismic study of the basin provides information about the subsurface structure and the possibility of any trap for the hydrocarbons.

Soil plays an important role for basin deposition and is also an indicator for paleo-climate, erosion, and depositional process. Soils are the product of weathering, which could be due to change of climate, different hydrological patterns, change of vegetation, and biological activities. If the sedimentation rate is high the soil residence time will be small. During erosion, there will be formation of platforms, terraces, and slopes. Transport of suspended load through river or wind is responsible for grain size deposition. When physical weathering dominates, the chemical weathering will be least. Presence of carbonate mineral indicates an absence of deep weathered soil. With a successive cycle of tectonic process, there was uplift, building of mountains, weathering, and renewal of sedimentation. During sedimentation, the sorting process plays an important role as it separates the different grain sizes in deposition. Clay associated with the sediments provides information about the source of sediments, rocks, and climate. The clay mineral distribution in oceans and seas provides information about the global climate; e.g., kaolinite is the product of a tropical area, whereas chlorite will be from a polar region. Sediments rich in illite are from a temperate to cold climate zone, while semectite is from a volcanic region and warmer climate zone.

The organic matter in the ocean has been input from a river system and also marine production of organic matter in the oceans. Tropical rain forests contribute larger amounts of organic matter to the oceans than the dry climate regions. The deposition of organic matter in a particular environment is related with hydrocarbon. Black or carbonaceous shale is the source rock for petroleum and contains more than 0.5% organic carbon. Carbonaceous shale is a fine-grained rock with low porosity and high permeability with clear laminations. The organic content varies from 0.5% to 20%. Carbonaceous shales that form in a reducing environment are the source rocks for hydrocarbon.

Basic factors controlling the deposition in basin are the size of the basin, geometry of the basin, production of biogenic matter in the basin, tectonic action, subsidence of basin floor, distribution of sediments, and sea level changes. The important factors for basin modeling are stratigraphic architecture, movements, and their accumulation. Such models are important for hydrocarbon exploration as they provide information about the source rock, reservoirs and cap rock. If the deposition of sediments in the sea from a river is through a lake, major sediments will be deposited in the lake, and a very small portion will reach the sea. Chemical sediments play an important role in basin deposition. Marine salt deposits rapidly, and you find a thick sequence in a short time. The Bengal fan is an example of rapid deposition of sediments from the Himalayan region through the Ganga and Brahmaputra rivers. The deposition is in such a large amount even at the equator that sediment coming through these rivers has been reported under a deep sea drilling program. Total sediments deposited in seas and oceans in the last 120 Ma after the breakup of Pangea derive 70% from continental slope and continental uplift and 17% from the shelves region (Emery and Uchupi, 1984).

Fine-grained carbonaceous shales are the source rocks for hydrocarbon. Carbonaceous shales convert the organic matter into kerogen of a different type based on maturity. The organic matter in buried sediments contains information of an organic compound produced by organisms. These organisms give the information about the source of organic matter and depositional environment. Study of biomarkers in oil is important to understand the source of hydrocarbon from plants or animals. In oil shale, total organic carbon (TOC) will be more than 20%, and mainly organic matter is converted into kerogen, which is not soluble in water. The preservation of organic matter in shale is a function of oxygen content in the water and rate of sediment deposition. Burial of organic matter with increasing temperature leads to conversion of organic matter into kerogens. Bitumen is the organic matter soluble in organic solvents. Kerogen has the potential to produce oil or gas, which may migrate to the reservoir rock. Kerogen has been classified into four groups. Liptinite type kerogen is the product of algal material after partial bacterial degradation, decomposition, condensation, and polymerization. These are rich in TOC and have a high Hydrogen/Carbon (H/C) ratio and finely laminated forms in lakes and lagoons. They are defined as type I kerogen. Exinite type kerogen or type II kerogen are from plant debris like pollen, spores, and leaf cuticles. They grow in lakes as well as in the ocean. With a high H/C ratio, they have good potential for generating oil and wet gas. Kerogens are the product of organic matter. They occur in lakes and marine sediments with limited potential for oil and condensate. The fourth type is inertinite type kerogen or type IV kerogen, which has a low H/C ratio and high maturity with no potential for hydrocarbons. To look at the evolution of organic matter to hydrocarbon, the first phase is the conversion of organic matter into insoluble kerogen, referred to as diagenesis. In next stage, kerogen converts into hydrocarbon, referred to as catagenesis. This is the main stage of hydrocarbon generation. The next stage is metagenesis where the source rock releases gas. At shallow depth, organic matter releases biogenic methane, like in a paddy field. In the mature stage, oil is generated, and in the late mature stage (temperature $> 130°C$), organic matter is cracked to form light hydrocarbons. The amount of oil and gas generated from kerogen depends on time period and temperature. Initially, oil accumulates in the intergranular space of the pore system, and after the saturation level, oil migrates toward a macropore system like sandstone.

Reference

Emery, K.O., Uchupi, E., 1984. The Geology of the Atlantic Ocean. Springer, New York, Heidelberg. 1050 p and map set.

Exploration Technique

CONTENTS

Shale Gas. http://dx.doi.org/10.1016/B978-0-12-809573-7.00005-6

Chapter 5.1

Shallow Seismic

A.M. Dayal

CSIR-NGRI, Hyderabad, India

5.1.1 INTRODUCTION

Shallow seismology is a geophysical tool for the exploration of mineral and hydrocarbon in subsurface structure. It uses the basic principles of seismology to understand the deep structure of Earth. In seismology, seismic waves are generated using either an explosive or vibrator. When we make any explosion in the subsurface it generates two types of waves. They are primary waves and secondary waves. They are also known as refracted and reflected waves. The primary wave moves through the solid, fluid, and gas medium, while secondary waves move only in solid structure. The velocity of primary waves is much higher compared to secondary waves. The primary wave has mechanical strength, and it can move the rock or the fluid through which it travels. Secondary wave moves slower than the primary wave. They can travel only through solids. To understand the deep structure of Earth, secondary waves are more important. Propagation of these waves provides information about the thickness of core, mantle, and crust.

Development of seismic technique as an exploration tool was an innovation by the petroleum industry. In seismic surveys, there is a source for the generation of seismic waves and receivers (geophones) for detection. It could be explosive or vibrating. These seismic waves travel through various rock formations, and their travel time depends on the density of rock type through which they are traveling. The velocity of waves and the density of rock it is traveling through can be defined by the following equation:

$$Z = V\rho$$

where V is the seismic wave velocity, ϱ (Greek rho) is the density of the rock, and Z is the seismic or acoustic impedance. The acoustic impedance will be different for different rock formations.

When a generated seismic wave travels in the Earth's interior, some of the wave will reflect and some will refract. The seismic reflection technique consists of generating seismic waves and measuring the time taken for the waves to travel from the source, reflect from an interface, and be detected by an array of receivers at the surface. The velocity of wave can be calculated by the time of generation of seismic wave and time of detection. The wave velocity will be different for basalt, granite, limestone, sandstone, and shale as the density of these rocks is not same.

The seismic wave generated from a source will be of two types. Some of the seismic waves that will be passing through the rock formation are called primary waves or refracted waves, and other seismic waves that will spread out as they cannot pass through media are called secondary waves or reflected waves. In a simple way in seismic exploration, time of generation of seismic wave and time of detection by various detectors placed at different distances are recorded. This data will allow us to determine velocity of the waves and indirectly the type of rock through which they has traveled. But it is not that simple, as the seismic waves will be generated in all directions from the source, and their propagation time will also depend upon the distances they have traveled. The primary waves can travel through solid, liquid, and gases, and these will be first waves to arrive at the detector. Compared to solids, liquid and gas will have less restriction for the propagation of primary waves. As the seismic wave moves away from the source, the intensity also decreases. If the seismic waves are traveling through homogeneous media the wave front of the seismic wave will be spherical in shape. Such waves are called plane waves. If a plane wave is traveling in a medium with seismic velocity, V1, when it passes through another medium, part of the wave will be reflected in the earlier media while the wave traveling in the new medium will have a different velocity, V2. Reflection seismology is more useful for the hydrocarbon industry.

5.1.2 THE REFLECTION EXPERIMENT

In a reflection experiment, a seismic wave is generated from the source, and it travels through different layers of the Earth and is reflected back, and the time is recorded by a geophone kept at a different distance but in a straight line. This will allow knowing the motion of these waves in the Earth's interior, which includes time of propagation and time of reflection. The seismic wave receiver or geophone converts this ground motion into an analog electrical signal. These recorded signals need processing for interpretation. The processing of these signals depends on the type of geological formation. In a complex region the processing will be very difficult for the data interpretation. Today, this seismic data is processed using high-speed computing systems.

5.1.3 REFLECTION AND TRANSMISSION AT NORMAL INCIDENCE

When a seismic wave produced at source travels through different media the energy and velocity of the seismic wave will be different in two different media. While passing from one medium to another medium, part of the wave will be passing through the medium and part of the wave will be reflected back to the earlier medium. The strength or amplitude of the reflected wave will be the amplitude of the incident wave in two different media.

For a seismic wave traveling at normal incidence, the reflection coefficient (R) will be

$$R = (Z_2 - Z_1) / (Z_2 + Z_1)$$

where Z_1 and Z_2 are the impedance of the first and second media, respectively.

For a seismic wave traveling at normal incidence, the amplitude of the transmission wave traveling through two different media, the incidence transmission coefficients (T) will be

$$T = 1 - R = (2Z_1) / (Z_2 + Z_1) \ (Shuey, 1985)$$

where R is the reflection coefficient, Z_1 and Z_2 two different medium.

5.1.4 REFLECTION AND TRANSMISSION AT NON-NORMAL INCIDENCE

But in the case of non-normal incidence the behavior of a transmitted wave and reflected wave will be entirely different and very complicated. In this case, there are number of reflection and transmission coefficients, R and T. This will generate a number of parameters for reflection waves as well as transmission waves. https://en.wikipedia.org/wiki/Reflection_seismology - cite_note-SheriffGeldart-1 If the angle of incidence is less than 30 degrees, Shuey (1985) suggested following relationship:

$$R(\theta) = R(0) + G\sin^2\theta$$

where R(θ) reflection coefficient at normal incidence with gradient G describes reflection behavior at intermediate offsets, and (θ) = angle of incidence.

5.1.5 INTERPRETATION OF REFLECTIONS

The travel time is the time taken by a seismic wave from the source to collection (geophone). If one knows the seismic wave velocity in the rock formation, the travel time can be used to know the depth of the rock for reflection. The travel time will be a two-way time: one for incidence and the other for the reflection and can be written as follows:

$T = 2d/V$

where d is the depth of the reflector, and V is the wave velocity in the rock. During seismic work in the field, a large number of such data is collected, which is later processed for the proper interpretation. By knowing the various seismic reflections from the analytical data, it is possible to find the various rock formations and the geological structure.

5.1.6 RAYLEIGH WAVE

A surface wave traveling near the surface of rock formations has been defined as a Rayleigh wave. The motion of these waves will be longitudinal as well as transverse, and their amplitude decreases as they traverse large distances. These waves are of low velocity and frequency, but amplitude is high and is defined as noise in the seismic data. Such waves are normally present in the seismic record and can be removed from the main seismic data by some filtration method.

Oscillatory motion parallel to the rock formations will be produce by the seismic wave when it refracts from the rock while traveling along it. Such oscillatory motions are disturbances to the upper medium and detected on the surface. The same phenomenon is utilized in seismic refraction. There are cases of multiple reflections also in seismic data recorded from one reflection. These are from water media and the air–water interface in marine seismic data. Such an interface is common in marine seismic data. While recording seismic data input from weather measurements, movements of airplanes or ships are recorded by a geophone, and such spurious seismic data has to be removed from the real seismic data. Such seismic data needs very specific data processing, so there is no contribution from spurious data in the final data interpretation.

5.1.7 APPLICATIONS

Reflection seismology is an excellent tool for geophysical exploration, specifically for hydrocarbon exploration. Use of reflection seismology can be classified into three groups based on depth. First is near surface, and such seismic data is useful while making any infrastructure, coal exploration, mineral exploration, or geothermal exploration. In such cases the maximum depth will be 2 km. For hydrocarbon exploration the seismic data is generated to understand the subsurface structure up to 10 km. For hydrocarbon exploration, very high resolution data is required to find the detailed structure in a subsurface area. To understand the interior of the Earth, seismic data up to 100 km is required to understand the thickness of crust and mantle.

5.1.8 SEISMIC DATA PROCESSING

The purpose of processing seismic data is to manipulate the data for various subsurface features. Seismic signals are converted to electronic signals (digital) using transducers. The digital data is processed and transmitted into waveforms. These waveforms are again processed and transmitted as electromagnetic waves that are received by other transducers in final form. Analog signal processing is for the signal received in analog form and needs digitation using various means. Digital signal processing is carried out using computers with a specific program. It uses different arithmetical formulations like Fourier's transformation, finite/infinite impulse response, and various filters.

Convolution is a calculation for two functions to produce a third function. It is a cross-co-relation and useful for statistics, probability, image processing, and various engineering applications. The inverse of convolution data processing is deconvolution data processing. A deconvolution is algorithm-based processing. An algorithm can be defined as the processing under a definite sequence of all computer operations. It can be used in space and time. An algorithm is used to define notion of decidability. Decidability is related to a decision problem. Decidable set concept is a logical system with computable functions. Theory is a set of formulas under logical consequence.

The deconvolution is used for signal and image processing for scientific and engineering fields. Deconvolution can be used to find the solution of a convolution equation with the following equation:

$$f * g = h$$

where f is the signal in which we are interested, h is the signal recorded, and g is the signal through noise generated by various means. It is necessary to know the g for data processing, and it is often done by statistical estimation. If ε is the noise in the data system, then that equation can be written as follows:

$$(f * g) + \varepsilon = h$$

If a signal is without any noise, the statistical estimate g and f may not be correct. To avoid such problems, inverse filtering is used. Deconvolution is computed using Fourier theorem on the recorded data h and transfer function on g. In the absence of noise the equation will be the following:

$F = H/G$

where F, G, and H are Fourier transform functions of f, g, and h.

The deconvolution concept is used in reflection seismology. In 1952, Enders Robinson discussed convolutional modal and reflection seismogram.

The common midpoint (CMP) is basically the propagation of seismic waves in the subsurface from a common point or source in all directions. In CMP data, amplitude is calculated by lowering the random noise, but we are also losing useful information between seismic amplitude and offset. Seismic data collected on land needs correction for the elevation between the shot and receiver. This correction is known as residual statics correction.

5.1.9 SEISMIC DATA INTERPRETATION

The interpretation of seismic data provides information about the subsurface that is useful for exploration and also engineering geophysics. In seismic data interpretation, 2D or 3D seismic data is used to know details about the geological subsurface structure. Such data will be useful to relate the other geophysical data and proper structure in the subsurface for coal exploration, hydrocarbon traps, or presence of geothermal fluids. Such data is also useful for the construction of any infrastructure like a dam, atomic power plant, or bridge. Besides the scientific and technical information, the subsurface geological information in the form of major fault systems will be useful to the community for any major natural damages like earthquake.

Reference

Shuey, R.T., 1985. A simplification of the Zoeppritz equations. Geophysics 50, 609–614.

Geochemical Exploration

A.M. Dayal
CSIR-NGRI, Hyderabad, India

5.2.1 INTRODUCTION

Geochemical properties of shale are important parameters to identify their production potential. The geochemical parameter includes content of organic carbon and thermal maturity of the shale. The geomorphology of the shale formations is equally important. Shale contains organic and inorganic carbon as total carbon. To quantify organic carbon, a combustion technique is used. To remove the inorganic carbon, a sample is treated with hydrochloric acid. For the type of kerogen and thermal maturity, Rock Eval pyrolysis analysis is carried out. The oxygen index (OI) is oxygen richness of the sample. Isotopic signature provides information about the source rock characterization. It could be biogenic, thermogenic, or a mixed source. This study is expected to provide sufficient geochemical data for the exploration of shale gas in the basins.

5.2.2 MAIN FACTORS CONTROLLING THE PRODUCTION OF SHALE GAS

1. high gas generation controlled by geochemical characteristic of shale,
2. high gas retention controlled by geochemistry and rock properties of shale source rocks,
3. source reservoir rocks for shale sequence vary within the shale sequence,
4. shale fractures also control the production of shale gas.

5.2.3 BASIC GEOCHEMICAL INFORMATION REQUIRED FOR SHALE GAS EXPLORATION

1. thermal maturity of shales containing different type of kerogens and organic content,
2. the lithological, petrophysical, and mineralogical character of shale that controls the fractures,
3. ratio of free gas versus adsorbed gas

Geochemistry of shales helps us understand the source rock characterization. For source rock characterization, the following information is required:

- organic richness
- thermal maturity
- type of kerogen

Organic richness is measured by a TOC analyzer in the shale and expressed as percent of rock. TOC percent in the shale sample is an indicator of the presence of hydrocarbon as given next:

TOC < 0.5% = no hydrocarbon
TOC 0.5–1% = some chances for hydrocarbon
TOC 1–2% = good chances of hydrocarbon
TOC > 25 = very good chances
TOC decreases with thermal maturity

There are four types of kerogen: Type I (oil prone), Type II (gas prone or oil prone), Type III (gas prone), and Type IV (no hydrocarbon) (Tables 5.2.1 and 5.2.2). Rock Eval pyrolysis and TOC for the source rock provide the following information:

S_1 = free hydrocarbon from C_1 to C_{23} thermally liberated from the rock sample at 300°C
S_2 = hydrocarbon cracked from kerogen or from C_{24} bitumen's in rock sample between 300 and 600°C
S_3 = organic CO_2 released form rock sample between 300 and 400°C
T_{max} = the temperature for the highest yield of S_2 hydrocarbons
Hydrocarbon index (HI) = $(S_2/TOC) \times 100$
Oxygen index (OI) = $(S_3/TOC) \times 100$
Production index (PI) = $S_1/(S_1 + S_2)$
S_1/TOC is the migration index

Calculation of Original TOC

$$TOC_{original} = TOC_{measured}/(1 - F \cdot dV_{TOC})$$

where F is the fraction of hydrocarbon generated = $(HI_{original} - HI_{measured})/HI_{original}$; dV_{TOC} = Max TOC convertible to kerogen type.

Table 5.2.1 Type of Kerogen Based on Element Ratio

Type of Kerogen	H/C Ratio	O/C Ratio	H/C Ratio	O/C Ratio
Type I	1.9–1.0	0.1–0.02	–	–
Type II	1.5–0.8	0.2–0.02	–	–
Type III	–	–	1.0–0.5	0.4–0.02
Type IV	–	–	0.6–0.1	0.3–0.01

Table 5.2.2 Classification of Kerogen and Their Potential Environment

Type of Kerogen	Maceral Composition	HI (mg HC/gm TOC)	Thermal Maturity (%)	Generated Hydrocarbon	Depositional Environment
Type I	Algal and amorphous organic matter	>700	0.6–0.9 0.9–1.3 1.3–2.6	Oil Gas condensate Gas generation	Highly anoxic/ lacustrine shallow marine lagoon environment
Type II	Amorphous organic matter and lipids	400–700	0.5–0.9 0.9–1.3 1.3–2.6	Oil Gas condensate Gas generation	Anoxic marine basin, lacustrine subtidal, supratidal environment
Type II/III	Mixed organic matter (amorphous organic matter and vitrinite)	200–400	0.6–1.0 1.0–1.3 1.3–2.6	Oil/gas cond. Gas cond. and gas Gas	Swamp and delta complex Low to medium anoxic lagoon
Type III	Mainly vitrinite	50–200	0.6–1.0 1.0–2.6	Gas/cond. Gas	Brackish water swamps
Type IV	Initernite	<50	No sources of hydrocarbons	Little gas	Oxic swamps and oxic marine basins

Type I = 62.5%, Type II = 48.2%, and Type III = 25.2%

Calculation of original HI = $HI_{original} = HI_{measured}/(1-F)$

Geochemical Modeling for Gas Generation and Retention in Shales

Free gas and adsorbed gas can provide better resource assessment of shale gas.

1. thermal extraction using gas chromatograph
2. free gas from gas chromatograph and isotope characterization

These data can be plotted in a Bernard diagram (Whiticar, 1994). $\delta^{13}C1$ versus C2+ can provide information about the characterization and origin of gas. $\delta^{13}C1$ versus $\delta H_{methane}$ can provide information about the characterization and origin of gas. $\delta^{13}C1$, $\delta^{13}C2$, $\delta^{13}C3$, and $\delta^{13}C4$ versus C1, C2, C3, and C4 can provide information about the maturity of thermogenic gas. It is important to know the content and volume of the gas in the shale formation. In a shale formation, gas is associated as free gas and also adsorbed gas. For the estimation of total gas in shale formations, the volume of gas with time can be measured in a core sample from a well. For an adsorbed gas, the sample is pulverized, and methane is measured with the function of time in a high-pressure condition. The gas adsorbed by a shale formation is presented in standard cubic feet per ton. Besides geochemistry, petrophysical and mineralogy are important parameters.

Reference

Whiticar, M.J., 1994. Correlation of natural gases with their sources. In: Magoon, L., Dow, W. (Eds.), The Petroleum System — from Source to Trap, vol. 60. AAPG Memoir, pp. 261–284.

Chapter 5.3

Petrophysics

A.M. Dayal
CSIR-NGRI, Hyderabad, India

5.3.1 INTRODUCTION

Gamma ray measurement is the most important part of any shale gas exploration because shale with high organic content has high gamma ray intensity. A gamma ray survey can be carried out in an unexplored sedimentary basin to get a broad picture about the presence of organic shale. The geochemical analysis of organic shale shows the high content of potassium, uranium, and thorium, which are high gamma ray material. A gamma ray survey for a shale formation of Cretaceous and Mesozoic age is not useful. For that purpose, another geophysical exploration like resistivity and porosity work is necessary. The association of natural gas in shale formations is as free gas and adsorbed gas. Some of the natural gas may be present in the form of kerogen in a shale formation.

5.3.2 POROSITY MEASUREMENT

Being a very fine-grained sediment the porosity of shale is very low. Porosity is the empty space in a shale formation that can work as a reservoir for natural gas, oil, and water. For the measurement of porosity of the rock, one has to calculate the total porosity of the rock, which is total volume of the rock divided by total pore volume. Organic-rich shale has higher porosity. Presence of gas in the organic shale reduces the hydrogen content. Organic-rich shales have higher density than sandstone or limestone reservoir rock. Porosity and permeability are also related to the mineral composition of the shale formation, distribution of organic matter in the shale formation, and thermal maturity of the associated organic matter.

5.3.3 MINERALOGICAL ANALYSIS

For the characterization of unconventional reservoirs, evaluation of shale formation mineralogy is required. Basic composition of shale is quartz, carbonates, feldspar, mica, pyrite, and phosphate. These materials are the weathering product of preexisting rock and are transported by wind and water and accumulate in a marine environment. Shale formations are not very homogeneous, as the environment of deposition will not be the same throughout geological time. X-ray diffraction (XRD) of a shale core is necessary to understand the mineralogy of the shale formation. Along with XRD, major, trace, and rare-earth element analysis is required in the shale under exploration. It has been observed that organic shales are enriched with certain heavy elements. It is important to understand the distribution of different minerals in the shale formation, and for this purpose, scanning electron microscope work is carried out. It will also allow to look for the microfractures in the shale samples, which will be useful while carrying out hydraulic fracturing. Mineral composition is necessary before the hydraulic fracturing work is to be carried out. Though being very fine grained, there is a very small volume between the layers, but it is good enough to store water, gas, or oil in these lattices. Shale is a combination of clay minerals and non-clay minerals. Major minerals in shale are kaolinite and illite with quartz and feldspar. The pore network in gas shales is different than the conventional gas reservoirs. The various porous media in gas shale are a nonorganic matrix, organic matter, natural fractures, and hydraulic fractures. Organic matter stores the free gas in a shale formation. Permeability of organic matter is very high, and this is the reason organic matter allows the migration of gas or oil in a shale formation. Movement of pore fluid in gas shale is through free gas movement, desorption, diffusion, and suction. It is important to study the mechanical properties (Young's modulus, Poisson's ratio, shear modulus, and comprehensive strength) of anisotropic shale for lengthy production of natural gas. The permeability of shale is based on its microstructure.

Some of the mechanical properties of shale formations are necessary to understand the stiffness and brittleness that help us know whether a fracture will remain open or collapse after hydraulic fracturing. Young's modulus and Poisson's ratio are the mechanical properties of the shale. These are related with lamination and orientation in shale. If there is large difference between vertical and horizontal Young's modulus in anisotropic rocks, the stress will be higher than isotropic rocks. Weak shale is not good for fracturing. The shale should have highly cohesive and unconfined comprehensive strength for hydraulic fracturing.

Shale Mineralogy

A.K. Varma

Indian School of Mines, Dhanbad, Jharkhand, India

5.4.1 INTRODUCTION

Shales occur widely, being almost 50% of all sedimentary rocks (Boggs, 2009) in the Earth's crust. These rocks are also known by various terminologies like siltstones, mudstones, mudrocks, claystone, clays, argillaceous materials, and shales. Sometimes words such as soil, weak rock, or soft rock are used loosely as synonyms for shale. Shales are fine-grained sedimentary rocks with a clay content in excess of ~40% and clay-sized particles along with clay minerals comprising 25% of total rock volume (Picard, 1953; Shaw and Weaver, 1965; Jones et al., 1989). Boggs, 2006 describes shales as fine-grained, siliciclastic sedimentary rocks that consist dominantly of silt-sized (1/16–1/256 mm) and clay-sized (<1/256 mm) particles. Shale is also referred to as a fine-grained, clastic rock that displays the property of being fissile. Illite, mixed layer illite/smectite, smectite, kaolinite, and chlorite are the dominant clay minerals in shales (Boles and Franks, 1979; Boggs, 2001; Day-Stirrat et al., 2010; Aplin and Macquaker, 2011). Shales dominantly contain admixtures of fine-grained quartz and clay minerals as well as other minerals, viz. feldspars, carbonate minerals, sulfide minerals, and oxide minerals (Yaalon, 1962; Vine and Tourtelot, 1970; O'Brien and Slatt, 1990; Slatt and Rodriguez, 2012).

Shales are considered one of the most problematic rock types for their applications in engineering domains (Farrokhrouz and Asef, 2013). Although they have been studied for many decades, shales are still a serious problem in engineering industries because of their sensitivity to mechanical, chemical, and thermal properties. The behavior of shales is delicate and complicated. Shales are defined in two ways: (1) general definition pointing toward physical properties based on microscopic properties and features (geological point of view) (2) specialized definition for engineering applications based on macroscopic properties (engineering viewpoint). The geological point of view about shales is more abundant than the engineering viewpoint. Potter et al. (2005) have recommended mudstone as fine-grained rocks and have used the term shale for fissile varieties. Here, the author has included both clayey rocks and all fine-grained siliciclastic rocks under shales.

5.4.2 MINERALOGY

Owing to their fine-grain size, shales require several other methods of investigation that are conventionally employed for sandstones and conglomerates. Interior structure and particle orientation of mudstones are studied by using a scanning electron microscope and the petrographic microscope. Mineralogy of the shale is studied through XRD and chemical composition using X-ray fluorescence (XRF) and inductive couple plasma–mass spectrometer. Boggs and Krinsley (2006) described these instrumental techniques in brief. Shales contain clay minerals, micas of very fine size, quartz, and feldspar dominantly as mineral constituents. However, organic matter contributes a little portion in ordinary shale to a reasonable amount in oil shale, gas shale, and mature hydrocarbon sources. XRD methods with an electron probe micro-analyzer and XRF can be used to investigate the quantitative composition of shales. Mineral composition of shales varies because of different geological aspects. As an example, quartz is the most dominant mineral in coarse-grained shales, while clay minerals concentrate dominantly in fine-grained shales (Boggs, 2009). Quartz constitutes about 20–30% of the average shale (Boggs, 2009). Quartz grains are mostly characterized by single crystals. Their size ranges from silt size to fine clay size. Single quartz crystals are mainly detrital grains produced by abrasion of larger grains or may be produced diagenetically during the transformation from smectite to illite or may be the recrystallized remains of silica-secreting organisms like diatoms and radiolaria.

Feldspar are less abundant than quartz. Generally feldspar consists of plagioclase and orthoclase. Zeolites, iron oxides, carbonates (calcite, dolomite, and siderite), sulfides (pyrite), sulfates, heavy minerals, etc., are present as minor elements in shales (Boggs, 2009). The most common carbonates in shales are calcite and are produced either by diagenesis or as detrital grains produced by carbonate-secreting organisms. However, two types of shales can be found: (1) shales bearing more than 50% clay minerals with little calcite mineralization and (2) shales with less than 50% clay characterized by more calcite mineralization (Varma and Panda, 2010). The average shale includes mainly 59% clay minerals, 20% quartz and chert, 8% feldspars, 7% carbonates, 3% iron oxide minerals, 1% organic matter, and 2% other minerals. Illite is the major mineral among the clay minerals. The physical and chemical properties of shales are chiefly controlled by the composition and structure of clay minerals (Yaalon, 1962). Pettijohn (1975) indicated that average shale is comprised of 33% clay minerals, 33% quartz, and 33% feldspars, carbonates, Fe minerals, and others minerals. According to Clarke and Washington (1924), it contains 25% clay, 22% quartz, and 30% feldspar. The feldspar amount of Clarke and Washington was considered excessively high since some of the potash may have been present in sericite. Krynine (1948) pointed out that the shales are a mechanical mixture, comprised of 50% silt (mainly quartz), 35% clay, and 15% authigenic minerals and cement. It appears that the grain size of fine-grained rock, i.e., shale, varies widely. The mineral composition of shales appears to control their chemical composition. In shales, SiO_2 is the most abundant chemical constituent, followed by Al_2O_3.

5.4.3 CLAY MINERALS

The definition of clay minerals adopted at aegis of C.I.P.E.A (Comite International pour l'Etude des Argiles; International Committee for the Study of Clays) in Brussels in July 1958, is as follows (Mackenzie, 1959): "Crystalline clay minerals are hydrated silicates with layer or chain lattices consisting of sheets of silica tetrahedral arranged in hexagonal form condensed with octahedral layers; they are usually of small particle size." Clay minerals are the principal constituents of any shale. Identification of different minerals through scanning electron microscope (SEM) was facilitated by comparing their characteristic morphologies with those shown in the SEM Petrology Atlas of Welton (1984). The mineral changes involved in the transformation of clay minerals into shales are associated mainly with dehydration (Lewis, 1924).

5.4.4 VARIOUS FACTORS CONTROLLING THE PROPERTIES OF CLAY MINERALS

The various factors controlling the properties of clay minerals or the attributes that must be identified to describe fully a clay material may be classified as follows (Grim, 1953):

1. clay mineral composition: relative abundance of all the clay mineral constituents,
2. non-clay mineral composition: relative abundance of non-clay minerals and their particle size distribution,
3. organic material: type and amount of organic material contained in the clay material. The organic material is generally present in two forms: (1) as discrete particles of wood, leaf matter, spores, etc., and (2) as organic molecules adsorbed on the surface of clay mineral particles,
4. exchangeable ions and soluble salts: some of the clay materials may have water soluble salts. The clay minerals and some organic matter present in clay materials have notable ion exchange capacity,
5. texture: particle size distribution of the constituent particles, shape, and orientation of particles in space with respect to each other and the forces appearing to bind the particles together.

The types of clay minerals includes these:

Kaolinites

This clay mineral group includes kaolinite, dickite, nacrite, and halloysite, of which kaolinite $[Al_2Si_2O_5(OH)_4]$ is the most common mineral. Often these clay minerals are grouped as kandites (Bell, 2009). Kaolin contain impurities (Murray, 1999). They are formed in warm humid climates due to excessive leaching and removal of Na, Ca, and K ions in solution during the chemical weathering process of orthoclase feldspar. So, the presence of kaolinite is a good indicator of paleo-environment and paleoclimate as it is the weathering product of a humid subtropical to tropical climate. Primary kaolinites have been altered in situ and generally retain their texture and form of parent rock. However, kaolin deposited in fresh or brackish water environments by sedimentary processes are referred to as secondary kaolin (Murray, 1999). Pure kaolin are nonabrasive, which is reflected by their hardness value of 1.5 in Moh's scale of hardness. Kaolinite is significant because of its primary dominance in china clay (Bell, 2009; Varma and Panda, 2010). Kaolin also exhibit hydrophilic properties. They get dispersed in water with a chemical dispersant (Murray, 1999).

Illites

These clays $[K_{1-1.5}Al_4Si_{7-6.5}Al_{1-15}O_{20}(OH)_4]$ comprise hydrous micas, phengite, brammalite, celadonite, and glauconite (Ross and Kerr, 1931; Ross and Hendricks 1945; Varma and Panda, 2010). They are formed by weathering of feldspars and some mica in temperate climates. Also, they are the alteration products of smectite clays during diagenetic stage. Illites are most common clay minerals in marine shales.

Smectite

This clay group includes beidellite, nontronite, volkonsite, and montmorillonite as di-octahedral members as well as saporlite, hectorite, stevensite, and sauconite as tri-octahedral members. They are formed from alteration of mafic igneous and metamorphic rocks enriched in Ca and Mg in temperate climates. These clay minerals pose characteristic swelling/shrinking properties due to weak linkage by cations. They expand by inclusion of water between layers and shrink during expulsion of water.

Montmorillonite

It $[(\frac{1}{2}Ca, Na)\ 0.7(Al, Fe, Mg)_4\ (Si, Al)_8O_{20}(OH)_4.nH_2O]$ is one of the good examples of the smectite group. Bentonite clay dominantly consists of the smectite group. The smectite minerals included in bentonites contribute significantly to their utility in industrial matter.

Vermiculites

This group is characterized by association of cations in the interlayer that results in basal spacing near to $14\,\text{Å}$. The strong linkage of water molecules with layer structure resists the expansion possibilities. They are primarily the alteration product of mica and chlorite. This group can be subdivided into di-octahedral vermiculite and tri-octahedral Ni-vermiculite and Jeffersite (Mackenzie, 1959).

Palygorskite and Sepiolite

Palygorskite represents hydrated Mg-Al silicate material. Although having similar structural features of sepiolite and palygorskite, the former has a slightly larger unit cell (Murray, 1999). Both of them contain chains of double silica tetrahedron consisting of Al and Mg. The tetrahedra show regular occurrence and give rise to channels through the atomic structure. These two minerals are classified under the chain lattices category of crystalline silicates (Mackenzie, 1959).

5.4.5 ORIGIN OF CLAY MINERALS

Clay minerals are mainly the weathering product of a variety of minerals in the Earth's crust. The transformation from parent material to clay is achieved by two processes: (1) alteration and (2) recrystallization. In the first process, particle size becomes finer than before. Alteration from muscovite to fine-grained muscovite occurs through this process. Due to intense weathering, the crystalline structure is broken and turned into a colloidal form that is accompanied with a loss of a part of the potassium and addition of silicon. The final product has an electronegative charge and less rigid crystal form characterized by 2:1 structure. The primary minerals also get decomposed and recrystallized into clay minerals. Recrystallization takes place after complete breakdown of parent minerals. The intensity of weathering is certainly higher than alteration procedure. Kaolinite formation with 1:1 structure from aluminum and silicon-bearing solution generated from chemical breakdown of primary minerals having 2:1 structure is a perfect example of the recrystallization process.

5.4.6 STAGES OF WEATHERING

Clays at the surface of the Earth are produced by the contact between rocks and water:

Water + Rock = Clay

The carbon dioxide gas dissolves to form carbonic acid, which disassociates in hydrogen and bicarbonate:

$$CO_2 + H_2O = H_2CO_3 = H^+ + HCO_3^-$$

Reaction with this acidic solution dissolves the potassium (K) ions and silica from the feldspar, which ultimately results in transformation of feldspar to kaolinite.

$$2KAlSi_3O_8 + 2H^+ + H_2O = 2K^+ + Al_2Si_2O_5(OH)_4 \quad + \quad 4SiO_2$$

(orthoclase)   (kaolinite)   (dissolved silica)

The micas, chlorite, and vermiculite are from mild weathering, while kaolinite and iron are from much more intense weathering. Intermediate weathering intensity results in formation of smectite.

5.4.7 STRUCTURE OF CLAY MINERALS

Clay minerals can be classified as phyllosilicate minerals because they have a leaf or plate-like structure. They contain a combination of $SiO_4^{(-4)}$ ionic groups and metallic cations. The phyllosilicate structure of clay minerals (Fig. 5.4.1) involves two horizontal sheets as follows:

Silica Tetrahedron
It is composed of one silicon and four oxygen atoms. An array of silica forms a tetrahedron.

Alumina Octahedron
Aluminum and/or magnesium ions form dominant cationic groups that are surrounded by six oxygen atoms or hydroxyl (OH) groups forming an octahedral coordination. The octahedral blocks join together horizontally to build

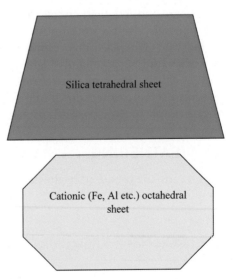

FIGURE 5.4.1
Tetrahedral (T)-octahedral (O) sheet.

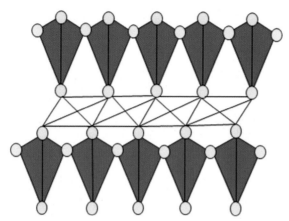

FIGURE 5.4.2
Tetrahedral (T)-octahedral (O)-tetrahedral (T) layers.

an octahedral sheet (O). An octahedral sheet contains aluminum as the dominant cation and a magnesium octahedral sheet; this requires an explanation. Two aluminum ions satisfy the configuration. These octahedral and tetrahedral sheets are tied together within the crystal structure by shared oxygen atoms into various layers in a different combination forming tetrahedral-octahedral-tetrahedral (TOT) layers (Fig. 5.4.2). The combination of these layers shows variation from one clay another and also controls the physicochemical properties of clay. Mainly, the TOT layers are combined in the following ways: diphormic (1:1), triphormic (2:1), and tetraphormic (2:1:1/2:2). All these structures are categorized under layer lattices of crystalline clay silicates (Mackenzie, 1959).

Diphormic (1:1) Clays

The layers are composed of silica and alumina of this group and include clay minerals that have the most prominent 1:1 structure. The TOT sheets are bounded by oxygen, silicon, and alumina. Hydrogen bonding developed between these layers results in a fixed structure with no expansion occurring between the layers when the clay is hydrated or wetted (Fig. 5.4.3). So, kaolinite neither absorbs water nor expands when coming into contact with water (Grim, 1953; Varma and Panda, 2010).

The structural layers of the diphormic clay type do not incorporate cations or water in between the crystal structures. The external surface area constitutes an effective surface of kaolinite. It experiences a little isomorphic substitution of Fe for Al and Al for Si in the crystal structure (Murray, 1999). Because of this little substitution the charge on the kaolinite surface is very low, which is responsible for the low cationic exchange. Due to smaller surface area and little isomorphic substitution, the adsorption of cation on the external surface

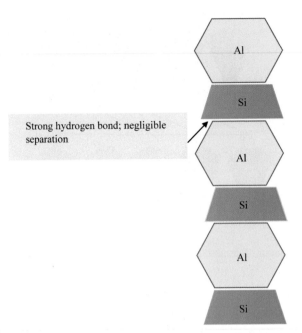

Strong hydrogen bond; negligible separation

FIGURE 5.4.3
Structure of kaolinites.

of kaolinite is very little (Murray, 1999). Strong bonding between the structural layers resists it from breaking down into thin fragments. The plasticity, swelling, shrinkage, and cohesion properties of kaolinites are very low.

Triphormic (2:1) Clays

A sandwiched octahedral sheet between two tetrahedral sheets depicts the characteristic of this clay group (Varma and Panda, 2010). This group can be described by smectite, vermiculite, and illite.

Expanding Minerals

This category includes the smectite group of minerals (Fig. 5.4.4) that swell during hydration. Montmorillonite shows its dominance within this clay group. The smectite crystal structure provides space between the layers. Swelling results as the internal surface of the clay crystal exceeds the external surface upon wetting. In montmorillonite, aluminum is replaced by magnesium in some octahedral cationic sites, which gives a rise in charge imbalance. There is also a replacement of silicon atoms by aluminum at tetrahedral sites, which again creates an imbalance in charge of 0.66 per unit cell (Grim, 1953). This imbalance in charge gives rise to high cation exchange efficiency, swelling, and shrinkage properties. The charge inequality developed because of these ionic substitutions is compensated by cationic exchange between unit layers and edges. Thus Na-montmorillonite

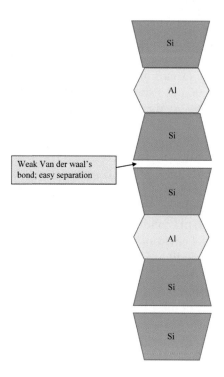

Weak Van der waal's bond; easy separation

FIGURE 5.4.4
Structure of smectite.

forms when the exchangeable cation is sodium and Ca-montmorillonite when the cation is predominantly calcium. Ca-montmorillonites have two water layers in the interlayer position with a basal spacing of 12.6 Å compared to Na-montmorillonite that have one water layer with a basal spacing of 15.4 Å (Murray, 1999). The water molecules and interlayer cation is a substituted compound of ethylene glycol and poly alcohols that will provide some useful organo-clay products (Murray, 1999). High charge, cation exchange capacity, and high surface area contribute to smectite having a high degree of absorbency for different types of materials (oil, water, chemicals, etc.).

Vermiculites also have octahedral sheets sandwiched between two tetrahedral sheets showing 2:1 type structure. They exhibit two types of cationic assemblages, i.e., di-octahedral and tri-octahedral (jefferisites). Silicon substitutes for aluminum in tetrahedral sites, which results in clay minerals. Magnesium as well as water molecules are strongly adsorbed in the interlayer spaces of the crystal structure, which serve as a bridge to hold the units together rather than separating them apart. This accounts for a lower degree of swelling in vermiculites than smectite. Thus, vermiculites are characterized with limited expansion of minerals that have the swelling tendencies greater than kaolinites but less than smectite.

The crystal size of vermiculites is bigger than that of smectite but smaller than that of kaolinites. The strong negative charge at tetrahedral sites of these clays causes cation exchange. Heating up to 1000°C results in exfoliation, a property that is characterized by expansion of vermiculites between 6 and 20 times its original volume (Varma and Panda, 2010). This property causes liberation of water bound between the interlayer spaces and expands the layers perpendicular to the cleavage plane (Varma and Panda, 2010).

Nonexpanding Minerals

Illites and micas are the typical minerals in this group. Illites, the main constituents of shales, are the weathering product of a temperate climate or the higher altitudes of a tropical climate. They are derived by the alteration of muscovite and feldspar under high pH condition (Varma and Panda, 2010). They are transported to a depositional environment (ocean) by river and wind. They are also 2:1 phyllosilicate minerals like micas but lack alkalis in the crystal structure along with less aluminum than silica. The interlayer cations like Ca, Mg, and K inhibit the entrance of the water molecules into the crystal structure, which accounts for nonexpansion property of these type of clay minerals (Fig. 5.4.5).

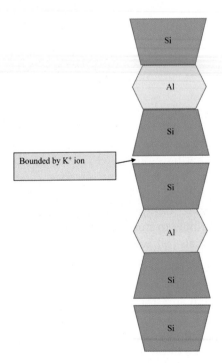

FIGURE 5.4.5
Structure of illites.

Fine-grained micas are another member of this nonexpanding clay group. Although they have a 2:1 crystal structure, the particles are much larger compared to smectite. Occupancy of aluminum of about 20% tetrahedral silicon sites results in sheets. To compensate the imbalance, monovalent spaces fit into the spaces of adjoining tetrahedral sheets. Therefore, potassium ions here serve as a binding agent that prevents expansion of the crystal, and fine-grained micas do not expand when wetted. Fine-grained micas show more intense hydration, swelling, shrinkage, and cation adsorption properties than kaolinites but lesser than smectite. Also, their surface area is one-eighth of that of smectite.

Tetraphormic (2:1:1/2:2) Clays

Chlorites represent this clay group. They are characterized by having iron-magnesium silicates with a low amount of aluminum present in the crystal structures. Alternation of 2:1 layers with a tri-octahedral cationic sheet gives rise to 2:1:1 or 2:2 layers. Therefore, the crystal structure of chlorites contains two silica tetrahedral sheets along with two magnesium-rich octahedral sheets. Chlorites show interstratification of vermiculites or smectite. They do not adsorb water molecules in the interlayer spaces, which points out their nonexpanding nature.

5.4.8 MINERALOGICAL INFLUENCE OF CLAY

Kaolinites show well-developed blocky crystals within the pore networks that reduce the porosity of the shale gas reservoir but have very small effects on the permeability. Due to its stability in acidic solutions, it occurs as detrital minerals in continental depositional environments and as authigenic clays in sands (Varma and Panda, 2010).

Illites can be both authigenic and detrital, of which the fibrous authigenic ones appear as fur-like jackets on the detrital counterpart. This arrangement builds a bridge over throat passages between pore networks in a tangled mass (Varma and Panda, 2010). Illitic cement causes a harmful effect on the permeability of shales. Illite shows its abundance in marine sands and occurs as authigenic in sands (Varma and Panda, 2010).

Montmorillonites, as discussed above, swell if wetted, which makes them very susceptible to formation damage if drilled with water-based mud (Varma and Panda, 2010). At the onset of production, water replaces oil, providing space for montmorillonites to expand and wipe out the permeability of the lower part of reservoir (Varma and Panda, 2010).

Depending on the source materials and history of diagenesis, kaolinites, illites, and montmorillonite may be found at shallow reservoirs. With increasing burial, alteration of montmorillonite and kaolinites forms illites. At this stage, breakdown of montmorillonite generates overpressure that may be related to hydrocarbon expulsion (Selly, 1998; Varma and Panda, 2010).

5.4.9 SIGNIFICANCE

Shale is considered a fissile rock with the addition of siliceous and calcareous concentration (Ingram, 1953; Pettijohn, 1975; Spears, 1976). Shale contains a large amount of quartz, feldspar, and carbonate with low Poisson's ratio, high Young's modulus, and high brittleness value (Ding et al., 2012). Shale lithology and mineral composition are the main intrinsic factors controlling fracture development in shale. Shale as a general lithologic terminology includes dark siliceous, calcareous, carbonaceous, ferruginous, sandy shale varieties, and oil shale (Li et al., 2007). The brittle nature of shale is the result of the presence of high contents of quartz, carbonate, and feldspar. Therefore, natural and induced fractures tend to be formed under external forces, and this fracture development is favorable for gas migration and accumulation. During artificial fracturing, silica-rich shale is more prone to fracture than clay-rich shale (Li et al., 2007; Pan et al., 2009). Siltstone, fine sandstone, or sandstone interbeds as well as the presence of open or incompletely filled natural fractures in shale can enhance the permeability of shale reservoirs (Li et al., 2009). The argillaceous shale with low rupture is more brittle in nature to produce fractures. This causes brittle shale to be the target to produce the preferred shale gas. Clay content affects elastic properties but not the maturity of the shales (Han et al., 1986). The abundance of organic matter is attributed to a high sea level at the beginning of shale deposition when nutrient rich upwelling had occurred. Burial of the siliceous biosome (such as radiolaria) resulted in high content of siliceous matter in the TOC rich shales (Li et al., 2009). The transformation of smectite or kaolinite to illite is well documented during diagenesis at the burial temperatures observed in gas shales (Bjolykke, 1998; Chermak and Rimstidt, 1990). The release of silica during the smectite-to-illite transformation and precipitation as microcrystalline quartz is particularly relevant as it can impact the physical characteristics of the shales (Aplin and Macquaker, 2011; Peltonen et al., 2009). Shale dominantly consists of various accessory minerals with a variety of clay for which its petrographic analysis is a matter of great challenge (Sondergeld et al., 2010).

X-ray powder diffraction is used to determine the primary mineralogy of whole rock samples, which is basically obtained by the Brown and Brindley (1984) method in which the predominant mineral presence is identified with its prominent peak, whereas the secondary minerals are difficult to identify due to the hindrance of peaks by the dominant peak. The variation in structure causes large reflection differences in intensity within the same mineral. This method is suited for clay-rich mineral composition of rock, as diagnostic reflection of clay is weaker compared to non-clay minerals (Moore and Reynold, 1997).

The pressure conditions of shale gas reservoirs may be influenced by clay content. Clays may have water of three types: normal pore water, structured water (bonded to the layer of montmorillonite clays/smectite), and constitutional water (North, 1984). After burial of illitic or kaolinitic clays, a single phase of water expulsion occurs because of compaction in the first 2 km of burial. When montmorillonite-rich mud is buried, the structured water is expelled during the collapse of the montmorillonite lattice as it changes to illite. It has been observed that the transformation of montmorillonite to illite indicates that this change occurred at an average temperature of some 100–110°C right in the middle of the oil generation window. The actual depth at which this point is attained varies with the geothermal gradient (North, 1984). By integrating geothermal gradient, depth, and the clay change point, it is possible to exhibit a fluid distribution model for the various shale gas reservoirs. Barker (1972) has put forward the idea that water and hydrocarbons may be attached to the clay lattice; hence the hydrocarbons may be detached from the clay surface when dehydration occurs. The precise physical and chemical process for expelling oil/hydrocarbons from the source rock is not clearly known. Clay dehydration may be one of the several causes of over pressure. Inhibition of normal compaction due to rapid sedimentation, and the formation of pore-filling cements, can also cause high pore pressures (Deshpande, 2008).

5.4.10 ADSORPTION OF GASES ON CLAY MINERAL SURFACES

Adsorption represents a surficial phenomenon where molecules of gases and liquids adhere onto the surface of the solids. Shale gas is adsorbed on the surfaces of both organic matter and inorganic clay minerals. However, adsorption of the gas by clays is an interesting phenomenon. The adsorption of CO_2 is dependent on the cationic exchange in clay crystal structures and pressure, whereas methane adsorption is almost entirely a function of clay surface area (Jin and Firoozabadi, 2013). Adsorption isotherms along with Monte Carlo simulation experiments exhibit that for CO_2 molecules, cation exchange causes significant intensification in CO_2 gas sorption within the nano-pore networks of clays, particularly at low pressure. However, in the absence of cationic exchange, the sorption of CO_2 largely increases with rising pressure and reaches its maximum because of dense packing of the gas molecules in the pores. The reason for the high adsorption of CO_2 in clay nano-pores may be explained by strong electrostatic interactions and correlations between the clay surfaces. On the contrary, methane sorption in clay nano-pores is reduced significantly by cationic exchange (Jin and Firoozabadi, 2013). The interaction between methane and clay atoms occurs principally due to non-electrostatic Lennard–Jones interaction (Jin and Firoozabadi, 2013). The methane adsorption reaches its maximum with an increase in pressure due to intense correlations between the two walls for small pores. CO_2 has a strong quadruple moment responsible for its adsorption on clay atoms, whereas the zero quadruple moment of methane resists its sorption considerably. However, in comparatively larger pores, an increase in pressure induces almost maximum methane adsorption, which indicates surface area affects sorption capacity of clays. Moreover, at the same time, adsorption of CO_2 is raised with raising bulk pressure due to long-range electrostatic interactions between clay minerals and CO_2 (Jin and Firoozabadi, 2013).

Natural kaolinite in an uncompacted state was found to provide the reversible Type II nitrogen isotherm (Gregg, 1968). Barrer and Macleod (1954) carried the physisorption experiments on natural montmorillonite. The isotherms for the relatively nonpolar molecules oxygen, nitrogen, and benzene were found to be similar in form (Cases et al., 1992). Each adsorption branch seems to have the same typical Type II character as the nitrogen isotherms on kaolinite. Many works have been carried on the sorption of water vapor on various forms of montmorillonite and vermiculite. Gregg and Packer (1954) got a set of unusual Type I isotherms, and all of them displayed low pressure hysteresis.

Pillared clays can be applied on Na-montmorillonite where ions are replaced by aluminum hydroxide that enters into interlayer position (Murray, 1999). In this case, pores are recreated by exchanging cations with inorganic ions (Rouquerol et al., 1999; Varma and Panda, 2010). After stabilizing by thermal treatment with removal of H_2O and OH groups, the nano-scale oxide pillars are introduced into interlayer spaces (Rouquerol et al., 1999). Increase in the molar mass of Al_{13} accounts for stabilization of pillared clays. The pillared structures can also be stabilized by controlling additional chemicals. Reaction of Al ions with Mg, Si, Ti, or Zr rich compounds forms co-polymers that also stabilize pillared structures (Vaughan, 1988a,b; Roquerol et al., 1999; Varma and Panda, 2010). These pillared clays can be widely employed for absorbent uses. They are expected to provide Type I nitrogen isotherms (Diano et al., 1994; Cool and Vansant, 1996; Varma and Panda, 2010).

Acknowledgement

Author expresses his thanks to Mr. Santanu Ghosh and Mr. Suresh Kumar Samad of Coal Geology and Organic Petrology Lab., Department of Applied Geology, Indian School of Mines, Dhanbad for their help in preparation of manuscript.

References

Aplin, A.C., Macquaker, J.H., 2011. Mudstone diversity: origin and implications for source, seal, and reservoir properties in petroleum systems. AAPG Bull. 95, 2031–2059.

Barker, C., 1972. Aquathermal pressurizing: role of temperature in development of abnormal pressure zones. AAPG Bull. 65, 2068–2071.

Barrer, R.M., Mcleod, D.M., 1954. Trans. Faraday Soc. 57, 980.

Bell, F.G., 2009. Engineering Geology. Butterworth-Heinemann. 592 pp.

Bjolykke, K., 1998. Clay mineral digenesis in sedimentary basins–a key to the prediction of rock properties. Examples from the North Sea basin. Clays Clay Min. 3, 15–34.

Boggs Jr., S., 2009. Petrology of Sedimentary Rocks, second ed. Cambridge University Press, Cambridge. 600 pp.

Boggs Jr., S., 2006. Principles of Sedimentology and Stratigraphy, fourth ed. Prentice-Hall, Upper Saddle River, New Jersey. 662 pp.

Boggs Jr., S., Krinsley, D., 2006. Application of Cathode Luminescence Imaging to the Study of Sedimentary Rocks. Cambridge University Press, Cambridge.

Boggs Jr., S., 2001. Principles of Sedimentology and Stratigraphy, third ed. Prentice-Hall, Upper Saddle River, New Jersey, pp. 88–130.

Boles, J.R., Franks, S.G., 1979. Clay diagenesis in Wilcox sandstones of Southwest Texas; implications of smectite diagenesis on sandstone cementation. J. Sediment. Res. 49, 55–70.

Brown, G., Brindley, G.W., 1984. X-ray diffraction procedures for clay mineral identification. In: Brindley, G.W., Brown, G. (Eds.), Crystal Structures of Clay Minerals and Their X-ray Identification, vol. 5. Mineral. Soc. Monogr., London, pp. 305–360.

Cases, J.M., Berend, I., Besson, G., Francosis, M., Uriot, J.P., Thomas, F., Poirier, J.E., 1992. Mechanism of adsorption and desorption of water vapour by homoionic montmorillonite. 1. The sodium-exchanged form. Langmuir 8, 2730–2739.

Chermak, J.A., Rimstidt, J.D., 1990. The hydrothermal transformation rate of kaolinite to muscovite/illite. Geochim. Cosmochim. Acta 54, 2979–2990.

Clarke, F.W., Washington, H.S., 1924. The composition of the earth's crust. U.S. Geol. Surv. Prof. Pap. 127 1–117.

Cool, P., Vansant, E.F., 1996. Preparation and characterization of zirconium pillared laponite and hectorite. Microporous Mater. 6, 27–36.

Day-Stirrat, R.J., Milliken, K.L., Dutton, S.P., Loucks, R.G., Hillier, S., Aplin, A.C., Schleicher, A.M., 2010. Open-system chemical behavior in deep Wilcox group mudstones, Texas Gulf Coast, USA. Mar. Pet. Geol. 27, 1804–1818.

Deshpande, V.P., 2008. General Screening Criteria for Shale Gas Reservoirs and Production Data Analysis of Barnett Shale (M.Sc. thesis). Graduate Studies of Texas A&M University.

Diano, W., Rubino, R., Sergio, M., 1994. Microporous Mater. 2, 179.

Ding, W., Li, C., Li, C., Xu, C., Jiu, K., Zeng, W., Wu, L., 2012. Fracture development in shale and its relationship to gas accumulation. Geosci. Front. 3, 97–105.

Farrokhrouz, M., Asef, M.R., 2013. Shale Engineering: Mechanics and Mechanisms. CRC Press/Balkema, Taylor and Francis Group, London. 288 pp.

Gregg, S.J., 1968. Surface chemical study of comminuted and compacted solids. Chem. Ind. 11, 611–617.

Gregg, S.J., Packer, R.K., 1954. The production of active solids by thermal decomposition. Part IV. Vermiculite. J. Chem. Soc. 3887–3893.

Grim, R.E., 1953. Clay Mineralogy. McGraw-Hill Book Co., Inc.. 384 pp.

Han, D., Nur, A., Morgan, D., 1986. Effects of porosity and clay content on wave velocities in sandstones. Geophysics 51, 2093–2107.

Ingram, R.L., 1953. Fissility of mudrocks. Bull. Geol. Soc. Am. 64, 869–878.

Jin, Z., Firoozabadi, A., 2013. Methane and carbon dioxide adsorption in clay-like slit pores by Monte Carlo simulations. Fluid Phase Equilib. 360, 456–465.

Jones, T.G.J., Hughes, T.L., Tomkins, P., 1989. The ion content and mineralogy of a North Sea cretaceous shale formation. Clays Clay Min. 24, 393–410.

Krynine, P.D., 1948. The Megascopic study and field classification of sedimentary rocks. J. Geol. 56, 130–165.

Lewis, J.V., 1924. Fissility of shale and its relations to petroleum. Bull. Geol. Soc. Am. 35, 557–590.

Li, X.J., Hu, S.Y., Cheng, K.M., 2007. Suggestions from the development of fractured shale gas in North America. Pet. Explor. Dev. 34, 392–400 (in Chinese with English abstract).

Li, X.J., Lu, Z.G., Dong, D.Z., Cheng, K.M., 2009. Geologic controls on accumulation of shale gas in North America. Nat. Gas. Ind. 29, 27–32 (in Chinese with English abstract).

Mackenzie, R.C., 1959. The classification and nomenclature of clay minerals. In: A Report of the Decisions Reached at a Meeting on the Classification and Nomenclature of Clay Minerals Held Under the Auspices of Comit International pour l'Etude des Argiles at Brussels in July 1958.

Moore, D.M., Reynolds Jr., R.C., 1997. X-ray Diffraction, Identification & Analysis of Clay Mineral. Oxford University Press, Oxford. 378 pp.

Murray, H.H., 1999. Applied clay mineralogy today and tomorrow. Clays Clay Min. 34, 39–49.

North, F.K., 1984. Migration of Oil and Natural Gas, Petroleum Geology. Allen & Unwin, London, pp. 231–233.

O'Brien, N.R., Slatt, R.M., 1990. Argillaceous Rock Atlas. Springer-Verlag, New York.

Pan, R.F., Wu, Y., Song, Z., 2009. Geochemical parameters for shale gas exploration and basic methods for well logging analysis. China Pet. Explor. 6-9, 28 (in Chinese with English abstract).

Peltonen, C., Marcussen, O., Bjorlykke, K., Jahren, J., 2009. Clay mineral diagenesis and quartz cementation in mudstones: the effects of smectite to illite reaction on rock properties. Mar. Pet. Geol. 26, 887–898.

Pettijohn, F.J., 1975. Sedimentary Rocks. Harper, New York. 348 pp.

Picard, M.D., 1953. Marlstone a misnomer as used in the Unita basin. Utah Bull. AAPG 37, 1075–1077.

Potter, P.E., Maynard, J.B., Depetris, P.J., 2005. Mud and Mudstones. Springer-Verlag, Berlin.

Rouquerol, F., Rouquerol, J., Singh, K., 1999. Adsorption by Clay, Pillared Layer Structures and Zeolites. Adsorption by Powders and Porous Solids. Academic Press, San Diego, London, pp. 355–375.

Ross, C.S., Hendricks, S.B., 1945. Minerals of the montmorillonite group. U.S. Geol. Surv. Prof. Pap. 205 B 23–79.

Ross, C.S., Kerr, P.F., 1931. The clay minerals and their identity. J. Sediment. Petrol. 1, 55–65.

Selly, R.C., 1998. Elements of Petroleum Geology, second ed. Academic Press, London, pp. 268–269.

Shaw, D.B., Weaver, C.E., 1965. The mineralogical composition of shales. J. Sediment. Petrol. 35, 213–222.

Slatt, R.M., Rodriguez, N.D., 2012. Comparative sequence stratigraphy and organic geochemistry of gas shales: commonality or coincidence? J. Nat. Gas Sci. Eng. 8, 68–84.

Sondergeld, C.H., Newsham, K.E., Comisky, J.T., Rice, M.C., Rai, C.S., 2010. Petrophysical Considerations in Evaluating and Producing Shale Gas Resources. SPE 131768.

Spears, D.A., 1976. The fissility of some carboniferous shales. Sedimentology 23, 721–725.

Varma, A.K., Panda, S., 2010. Role of clay minerals in shale gas exploration-an evaluation. In: Varma, A.K., Dubey, R.K., Sarkar, B.C., Saxena, V.K. (Eds.), Geological and Technological Facets of CBM, Shale Gas, Energy Resources and CO_2 Sequestration. Allied Publishers Private Limited, New Delhi, pp. 99–107.

Vaughan, D.E.W., 1988a. Catalysis Today. Elsevier, Amsterdam. 187 pp.

Vaughan, D.E.W., 1988b. Pillared clays – a historical perspective. Catal. Today 2, 187–198.

Vine, J.D., Tourtelot, E.B., 1970. Geochemistry of black shale deposits; a summary report. Econ. Geol. 65, 253–272.

Welton, J.E., 1984. SEM Petrology Atlas. American Association of Petroleum Geologists, Methods in Exploratoin Series, 4.

Yaalon, D., 1962. Mineral composition of the average shale. Clays Clay Min. 5, 31–36.

CHAPTER 6

Hydraulic Fracturing

A.M. Dayal

CSIR-NGRI, Hyderabad, India

CONTENTS

6.1 INTRODUCTION

In carbonaceous shale the gas is stored between the lattices as free and adsorbed gas. It is possible to extract this gas from shale formation using hydraulic fracturing. In hydraulic fracturing, water with sand or ceramic balls and some chemicals as proponent are injected with very high pressure to create fracturing in these layers. Hydraulic fracturing had been used earlier for conventional hydrocarbon exploration and exploitation. But now hydraulic fracturing is a completely new technique being used for the shale gas industry. In the United States the development of the shale gas industry is just 1.5 decades old. With increasing global oil prices and limited reserves, it was necessary to find a new

Shale Gas. http://dx.doi.org/10.1016/B978-0-12-809573-7.00006-8

source of energy to replace existing conventional hydrocarbon energy, which happened with the coordination of industry and academia in the United States through horizontal drilling and hydraulic fracturing at the beginning of this century. Now, in the United States a large number of wells are hydraulically fractured to extract shale gas. With large-scale production of shale gas by 2014, the United States stopped importing gas from Canada. As they have increased the production of shale oil, the import of oil has been also stopped, and the United States has a more than 2000-billion-barrel excess of oil. Along with shale gas, they also developed coal-bed methane.

Shale is very fine-grained, laminated rock with very low permeability. To extract the gas preserved in these laminated rock formations the only way is to frack these rocks. In fracking, interconnected pores will allow the movement of associated gas to migration. In hydraulic fracturing, methane, which is a lighter gas, is easily replaced by water. Permeability of the rock can be defined as the ability to transmit fluid within the rock formation. Hydraulic fracturing increases the pathways for gas flow in a shale formation by several orders of magnitude.

Water is the major requirement for shale gas exploration as hydraulic fracturing; for one well, 2–3 million gallons of water is required. But compared to the requirement for one metro city per day, this is a very small amount. There is a large number of sources of water for this operation like surface water, groundwater, excess rain water, or treated sewage/industrial water. In hydraulic fracturing, 60% of the water is being used, while 40% of the water comes back as flowback water that needs treatment for reuse or disposal to surface water streams. For the shale gas industry, water management is one of the most important processes. At the drilling pad, construction of few lined reservoirs are required to store fresh water, water for hydraulic fracturing, flowback water, and treated water. During rains, water can be collected in the reservoir as per the yearly requirement, and later it can be disposed to surface stream after necessary treatment. Use of flowback water after necessary treatment is another way to reduce the fresh water requirement. Sometimes, flowback water is very bad, and it is difficult to clean as per the requirement; in such cases, water is injected back into deep wells. It is important to strictly follow the guidelines for water management, which were not the part of conventional hydrocarbon exploration.

6.2 HYDRAULIC FRACTURING

It is important to plan all the features related with hydraulic fracturing, which include source of water, water disposal, and safety measurements. A logbook is necessary to write all the actions taken, which includes procedures, operations, and any emergency operations. It is important to follow all the safety rules and regulations specified for the industry, which includes all permissions from government agencies and also from the local civic community as people will be affected. The well pad should be away from the living beings to avoid any disturbances to their regular life. For the operator, it is advisable to make their own road for the transport of their vehicles, and enough precautions should be taken to avoid any water contamination or large-scale emissions to the environment from the various equipment operating at the well pad.

Very good planning is required for hydraulic fracturing. It is important to ensure that during hydraulic fracturing shallow water aquifers are protected from any contamination from the fracking fluid or any emission in the environment that could be hazardous to the community or other living beings. In case the well pad is in the forest area, it is important to ensure that there is no impact to animal life with the process being carried out for hydraulic fracturing, which includes water body and high-decibel noise production. In the case of a conventional hydrocarbon bearing zone the wells in operation should be protected.

There are a few important components to be addressed before hydraulic fracturing is carried out. Some of these for horizontal drilling and the stimulation of the entire process for hydraulic fracturing are the amount of water to be used for fracturing, amount of proponents required for this fracking, the chemicals to be mixed with fracturing fluid, the volume in subsurface for hydraulic fracturing, pressure of injected fluid for hydraulic fracturing as it depends on the petrology of the shale formation, and the volume of the area to be fractured. For proppant, either sand or ceramic balls are used. With high-pressure hydraulic fracturing, it is possible to create minor crevices in the upper rock formation also, and these fine crevices may allow the movement of fracturing fluid or trapped hydrocarbon toward the bore well. To avoid such events, it is important to monitor the entire hydraulic fracturing process, which includes induced seismicity due to fracturing and emission of any gases from the well bore at high pressure to avoid any fire accident.

In hydraulic fracturing, fracking fluid is injected through main vertical well and then toward the horizontal wells in high-pressure pipe, and fracking is carried out at predetermined places in many directions. High-pressure fluid will open a number of laminar fractures in the target shale formation through perforations, and the volume of the shale formation will increase. This will induce seismicity at a very low level, which may not even be recorded at the surface level. Hydraulic fracturing of a shale formation is also known in industrial terminology as "breaking down." The opening of shale laminations will depend on the physical properties of the shale formation, amount of fluid being pumped for fracturing, and viscosity of the fracturing fluid.

Shale gas exploration and exploitation is a multiphase process. In the first phase, a shallow seismic survey is carried out to get an idea about the extent and thickness of shale formations in the basin. After necessary geochemical, geomechanical, and petrophysical analyses, the area is selected for hydraulic fracturing. The next phase is hydraulic fracturing of the well for the exploitation. During drilling the cementing of pipe and a strong lining is required to avoid any contamination to rock along the well bore. High-pressure lining work is necessary for the entire vertical and horizontal well. Normally the well bore at start will be 19 inches in diameter, and as we go down the diameter of the well bore decreases, and finally it will be 4–6 inches in diameter. In the top casing, cementing work is also carried out along with steel pipe. The entire well bore should be leak proof, and before hydraulic fracturing, perforating is carried out at a definite interval along the pipe.

Drilling of well itself is the major process to be carried out at the drill pad. Drilling fluid is used for the lubrication of drilling assembly. It also helps in removing the formation cuttings. Drilling fluid is the mixture of water, bentonite clay, and some additive for the viscosity of the fluid, but the composition of drilling fluid will be different for each type of geological formation. For horizontal drilling, initially vertical drilling is carried out till reaching the shale formation, and then horizontal drilling is carried out in the shale formation up to 5–7 km in multiple directions in different layers.

While casing work of the well is carried out, it is important to know the geological formations and their lithostratigraphy. In case of the presence of any fracture, which is a weak zone in the rock formation, casing at such point has to be very specific to avoid any leakage from such geological formations. Protection of drinking water at shallow water aquifers is most important for the community or in case of forest the animals using surface water for their routine requirement.

Safety of the people working with the shale gas project is of most importance. All the workers should be provided required safety material, and it is necessary to monitor that they are using all the safety material given to them for various operations at drilling pad, including vehicle operators, operators on the drilling rig, persons at a generator, and persons handling fracturing fluid and some of the hazardous chemicals. It is necessary to follow the procedures for every processes, and it is better that there are printed safety banners all over the operation area to remind the person about their safety. An onsite preliminary medical facility should be available at the operating area. For safety reasons the operation area should not be congested, and enough space should be available for the movement of vehicle, people at the well pad, and storage of various material. Periodically, all the people working at the well pad should be provided safety training for any accident that might occur at the well pad.

It is common to add some chemicals in the fracking fluid for various purposes. While adding proppant and other chemicals, there should be a well-documented procedure, and it should be strictly followed. It is always better to avoid storage of hazardous chemical at the well pad. For safety reasons the mixing of chemicals and their blending should be carried out in the presence of a senior, well-experienced operator. The operator should know about the chemicals and safety requirements while handling it.

6.3 HAZARDS

In shale gas exploration and exploitation, there are many potential hazards starting from drilling to well completion. With flowback water, there is a large amount of debris with very high pressure, and an experienced worker is required for the handling. The debris includes solids, liquids, and gas in various forms. It is necessary to handle the flowback water very carefully, initially reducing the pressure and separating all mix material with proper safety measurements. There could be flammable gas, which needs to be released into the atmosphere. As the flowback water is at high pressure, it is important to make a built-in provision for reducing pressure and separation of solids from liquid and release of gases into the environment. Monitoring of flowback water is necessary to avoid any accident from flammable gases in the high-pressure flowback water system. For safety reasons, it is important to check all the equipment before use, and regular maintenance should be carried out and maintained in a logbook. The high-pressure fluid line should have a minimum number of manifolds to avoid building pressure and thereby the possibility of an accident.

6.4 FRACTURING FLUID

In fracturing fluid, 90% is fresh water with 9% sand or ceramic balls as proppant and 1% chemical for various purposes, which includes viscosity of frack water and to avoid corrosion in the pipe line. Proppant is required to keep the fracture in a shale formation open after the return of flowback water. This will allow the release of free gas from the shale formation.

Earlier, sand was used as a proppant, but now ceramic balls are preferred based on their performance. Besides proppant, some chemicals are required, which are less than 1% by volume. These chemicals could be acids, biocides, pH stabilizers, iron inhibitors, and guar gum. Many chemicals are common products and part of household items. But there are few chemicals that are not safe, though they are added into the fluid in a very small amount, and they are hazardous to the health of living beings (Tables 6.1 and 6.2). These chemicals

Table 6.1 List of Chemicals Used for Hydraulic Fracturing

Product Function	Chemical Used
Proppant	Sand, ceramic balls
Acids	Hydrochloric acid, hydrofluoric acid
pH adjustment	Acid, bases
Biocides	Aldehydes, ammonia compounds
Friction reducers	Methanol, ethylene glycol
Gelling agents, polymers	Guar gum, polysaccharides, polyacrylamides
Cross-linkers	Sodium tetraborate
Breakers	Magnesium, calcium peroxide Magnesium oxide
Iron control agents	Citric acid, acetic acid
Scale inhibitor	Sodium polycarboxylate, phosphoric acid, salt
Clay stabilizers	Choline chloride, tetramethyl ammonium, and sodium chloride
Corrosion control	Formic acid, acetaldehyde
Surfactants	Lauryl sulfate, alcohols,

Table 6.2 List of Carcinogenic Chemicals

Benzene	It is carcinogenic, causes anemia, and reduces white blood cell counts. Long-term exposure may develop blood disorders, leukemia, and reproductive disorders
Toluene	Long-term exposure may affect nervous system
Xylenes	Short-term exposure may cause irritation of nose and throat, nausea, vomiting, gastric irritation, and neurological effects. Long-term exposure affects the nervous system
Methanol	Exposure can result in blurred vision, headache, dizziness, and nausea
Naphthalene	May cause abdominal pain, nausea, vomiting, and fever, low blood pressure or jaundice
Formaldehyde	Long-term exposure may cause lung and throat cancer
Acrylamide	Short-term exposure may cause damage to the nervous system
Ethylene glycol	At high exposures, may affect the central nervous system, heart, and kidneys

are used in industry. The operator should also avoid using such chemicals, as later for the safe disposal of flowback water, it may be difficult to remove these chemicals from the fluid mixture. Composition of fracturing fluid will not be the same for all the wells as it is based on the type of formation and their chemical composition and mechanical strength. Based on requirements the chemicals are added into the fracturing fluid.

6.5 WATER MANAGEMENT

Water management is the most important part of a shale gas play. This includes fresh water for fracturing, flowback water from the fracturing well, treatment of flowback water, reuse of treated water, and safe disposal of used water after proper treatment as per the guidelines for the clean water system. There are various water sources for fracturing fluid. In some cases, surface water is available in a good amount, and it can be used for hydraulic fracturing. Storage of excess rain water in a reservoir is another option that does not disturb the natural water system. In dry areas, water has to be transported from the nearest surface water zone, which is an expensive process. The most difficult part is the disposal of flowback water as it contains a large quantity of impurities, and some of dissolved elements cannot be separate out, and it is difficult to dispose such water. There are a few cases when arsenic, uranium, and thorium from a shale formation came in the flowback water, and it is very difficult to remove these elements prior to disposal.

The hydraulic fracturing process is designed according to the shale gas reservoir condition, characteristics of associated shale, amount of water required, and time frame for the operation. For a particular shale formation, simulations and designs of the fracturing process is necessary. The composition of fracturing fluid will be based on the shale formation to be fracked, which includes the amount of fracture fluid pressure and time duration to complete the process. Once the fracturing process is completed, low-viscosity fluid is circulated for cleanup of the system.

The last step for well completion is disposal of flowback water after necessary treatment into either surface water or to be injected into abandoned deep wells. Disposal of water needs enough care about the amount of water, its composition, and monitoring by a third party, as well as the local civic authority if necessary. The shallow water aquifers have to be protected from any contamination form the chemicals used in fracturing water. The association of any gas with flowback water should be released into the environment per the guidelines. All the muddy material needs to be separated, and it will also have lot of chemical additives that may contaminate groundwater; hence, such material needs a different type of processing. It can be mixed with the surface soil, but in such a way that the concentration of contaminated material is highly diluted and is no longer harmful for any use including agriculture.

For shale gas exploration, a large amount of water is required for hydraulic fracturing. Besides hydraulic fracturing, a small quantity of water is also required while drilling the large number of wells in the play area. In the play area, one has to see the various sources of water, and the selection should be done in such a way that transportation of water is at a minimum and at the same time the requirements of the local community are not being affected. Water used from any source needs permission from the local civic authority, giving all the details about the required quantity of water till well completion and mode of transport and safe disposal of flowback water. Even storage of excess rain water in the properly lined reservoir needs permission to avoid any untoward incidence for the industry as well as the people staying in the nearby area. The local community, civic authority, and people from the industry should be part of the water management committee, and they should be responsible for the sources of water, storage in reservoir with safety requirement, and in case of transportation, the road infrastructure facility in the area.

Normally, it is the practice to fracture a number of hydraulic fractures from one well pad. But in case of a greater number of well pads, it is important to assess the total water requirement, and the hydraulic fracturing should be carried out in a sequential manner. If the well pads are not very far, a basic facility for water reservoir, water treatment for flowback water, and disposal of treated water should be carried out from the main well pad. This will bring down the operation cost and cause the least disturbance to the civic authority and local community.

There are number of water resources and the amount of water available in these resources depends on the topography of the area. Surface water in the form of rivers and lakes is available in large amounts in rain-fed areas. This water can be exploited for hydraulic fracturing with proper planning. The next source of water is groundwater, but it depends on the recharging of the shallow water aquifers. In many places, this water can be used for hydraulic fracturing. In an urban area, there is a water supply available from municipal government, though this water may not be available in a large quantity, but for drilling work, such sources can be tapped. In a large metro city, a good quantity of treated water from industry and sewage is available that can be used for hydraulic fracturing. But being treated water, it needs to be mixed with fresh water for industrial use. In a power plant, a large quantity of water is available that can be tapped for such uses. Flowback water after necessary treatment can be used but has limited options.

Selection of a water source depends on the quantity and quality of water requirements and available water at the resource. The best practice will be water from a power plant, treated industrial waste water, treated flowback water, or even saline water, as this water will not affect the community and if the civic authority agrees for the use of such water with the condition of safe disposal. All these water sources are not available at one place, and the option is based on the location of the well pad.

Use of surface water is one of the easiest options, but large-scale use may not be available. For surface water, there are many users, which include the local community, agriculture, animals, hydropower plants, and thermal power plants. Use of this water in a large quantity will have severe impact on these users. Sometimes, there are lakes, but water that is not potable can be used for hydraulic fracturing with prior permission from the local civic authority. In a few countries, groundwater rights are with the landowner, and he can allow using water from his land for the industrial purpose. Withdrawal of water from a surface stream should not affect the fish nor aquatic life in the rivers, stream, or lakes. In case the civic authority and local community gives permission to use surface water, there should be monitoring by a third party to keep the account of the quantity of water being used daily, and industry should make a log book for such operations, which may be required by the monitoring authority. In case water is being transported from some distant water body, the operator requires permission from the local civic authority for laying the pipeline or using infrastructure like roads for transportation or construction of a canal. The operator may be asked to maintain the infrastructure facility being use for various purposes. If the river is not perennial, then water can be collected only during the periods of high flow. Based on the play area, the operator has to plan for the water sources. Rain water harvesting could be another approach in a low lying area under the guidelines from the local civic authority.

Groundwater is also one of the important sources in a good rainfall area. In fact, water from the groundwater can be collected in different time intervals, giving enough charging time. The use of groundwater in a large quantity requires civic authority permission. In some places, groundwater has very high TDS and is not potable; such water can be used for hydraulic fracturing. Withdrawal of groundwater in a large quantity from a deep well may have some impact to shallow water aquifers, and in such cases, the operator should not be allowed to withdraw water from groundwater.

Water supply from a municipal corporation can be used, but there will be limitations as the water requirement for the shale gas industry and the local community has to be balanced, and in a drought season, this option will not be available. Produced water and flowback water can be treated and reused for fracturing, depending on the quality of the water. Produced water is in contact with formation rock and will be highly turbid. The salinity and TDS of the formation and flowback water will vary by geology of the basin. Presence of hydrocarbon, suspended solids, soluble organics, iron, calcium, magnesium, boron, silicate, and trace constituents will be present in the flowback/formation water. Some coal-bed methane operations may also have discharge water that is appropriate for hydraulic fracturing use. In the case of use of treated water, some fresh water mixing is essential. Handling of frack fluid before hydraulic fracture and after the completion of hydraulic fracturing is itself a complete operation under water management. Fresh water as well as flowback water and treated water are stored in a lined reservoir near the well pad. For hydraulic fracturing, fracking fluid is mixed with proppants, chemicals, and fresh water and/or treated water. Handling of this fluid needs extreme care to avoid any wrong mixing or leakage of chemicals. After the completion of hydraulic fracturing, flowback water mixed with formation water needs treatment before reuse or surface disposal. For the safety of shallow water aquifers, all the water stored in the lined reservoir at the well pad needs extra precautions. Any overflow while preparing fracking fluid should be avoided. Return of flowback water initially will come in a large quantity, and in a very short time, a small quantity will continue for a few months.

For safety reasons, storage of chemicals in a large quantity should be away from the main well pad. These chemicals should be transported in a measured amount to be mixed or blended with fracking fluid, and this should be a stage at the start of hydraulic fracturing. This action avoids any pilferage of chemical at the well pad aw well as contamination to shallow water aquifers. All the unwanted material should be removed from the well pad zone. In the United States, flowback water is not treated as hazardous fluid, but it comes under all safety regulation procedures. Any spillage on the ground should be removed and disposed as per the guidelines. All the additives used in fracking fluid should be clearly mentioned in an operating log book. For the operators, it is required to provide all the information about their water management and storage operations at the site. This includes design, capacity, and number of storage reservoir for various purposes. All reservoirs should have a large display board and a physical barrier for every type of fluid storage to avoid any accidents.

All lined storage reservoirs for various fluids should be constructed per the written guidelines from the government issued for the exploration of shale gas. Volume of the storage reservoir should be 25% larger than the amount of fluid to be stored to avoid any overflow from the reservoir. The flowback reservoir should be large as the quantity of water coming back as flowback will not be known earlier. The lining of the reservoir is based on the type of fluid that will be stored and the duration of storage to prevent any leakage from the lining material. The flowback reservoir with long-time storage should be away from the fresh water reservoir. All storage reservoirs should be at aground level to avoid any downfall spillage. For the construction of any infrastructure facility like a water reservoir, it is important to study the topography, geology, and any fresh water bodies like wetlands, lakes, ponds, or streams.

In some cases due to topography of the area, fluid is stored in steel tanks. In such cases, all the steel tanks should be constructed under the guidelines issued by governing agencies, and monitoring of such storage vessels is periodically necessary. For the storage of flowback water, steel tanks should be painted with a required material to avoid any corrosion or other damage to the steel body.

6.6 TRANSPORTATION

Transportation of various materials like sand, water, chemicals, power equipment, drilling rigs, and temporary residents of the well pad is another part of the game in a shale gas play. For the transportation of all these materials, initial permission is required from the civic authority. Water is transported in tankers or via pipelines. In the case of a nearby urban area, the civic authority may allow the transportation only during night hours or lean period of traffic movement. In case of transporting a large amount of water, if the topography allows, it is better to transport water through a pipeline. In case of the use of groundwater, transportation of the water part can be reduced, and this should be worked out in consultation with the local community and civic authority. It provides close access for the operator to a water source, and it adds improvements to the property that benefit the landowner.

6.7 FLUID DISPOSAL

After the well completion, all the temporary structures including fluid reservoir for storage of flowback water, treated water, etc., must be removed. Water used in the hydraulic fracturing process can be disposed by injecting into deep wells under a permitted procedure. Flowback water can be treated per the guidelines and disposed with surface water or it can be reused for another fracking operation. Injection of water into a deep well is possible only when such wells are available. In the absence of deep wells, water can be treated per the required procedure and can be disposed with surface water. For any shale gas operator, it is necessary to make baseline data for water samples in the shale play area. While disposing the various fluid after necessary treatment, this baseline data is very important to look for any contamination. Baseline data include TDS, concentration of various salts, total suspended solids, and organic content. Injection of flowback fluids in a deep well is the most environmentally friendly procedure for the fluid disposal. Sometimes, additional deep wells are drilled for the disposal of flowback fluid.

In water scarce zone, it is advisable to treat the water as per issued guidelines for reuse by the community for particular purposes. Recycling of flowback water after necessary treatment for hydraulic fracturing is another solution for disposal and is being carried out in many shale gas play areas. Reuse of water depends on the degree of treatment and quantity of fresh water mixing. In a limited water supply zone, this procedure is quite useful. Today, with reverse osmosis and membrane technology, even highly saline water can be used as potable water. Distillation is another process for water purification, but it is an expensive process. Today's solar energy can be used for distillation, which will be a bit cheaper option. Technology advances are making it more economical to treat these fluids with better results in water quality. The treatment of these fluids may greatly enhance the quantity of acceptable, reusable fluids and provide more options for ultimate disposal. For any water treatment processes the performance and associated costs are key for the industry.

6.8 RISKS OF HYDRAULIC FRACTURING

In shale gas exploration and exploitation, hydraulic fracturing of shale formations has many risk-oriented processes. Use of fresh water in a large quantity, addition of chemicals, disposal of flowback water, emission of various gases from flowback water, and disposal of solid material are the measured risks in shale gas exploration. The induced seismicity during hydraulic fracturing is also one of the risk factors for the contamination of shallow water aquifers. While injecting used water into deep wells, a fair amount of chemicals associated with the fluid are mixed with groundwater and contaminate shallow water aquifers over a long time. The risk to groundwater is major risk in shale gas play.

There is a feeling among the community and environmental groups that hydraulic fracturing will contaminate the shallow water aquifers through the fracturing fluid as the induced seismicity will open a number of lineaments and minor fault systems. And the high-pressure fracking fluid may impact the groundwater system. There are emissions of gases during hydraulic fracturing, and during hydraulic fracturing, it is possible to release the associated gases to the environment through these small lineaments. In the case of coal mining, such incidents have been reported, but coal mining is carried out at shallow depth compared to shale gas exploration and exploitation. To avoid any such impact to groundwater, it is important to study the geology of the area well, particularly the structural geology. In the case of minor faulting or structurally disturbed formations, it is advisable to study the structure well before taking up the site for hydraulic fracturing. It is always possible that saline water or associated gases might migrate upward and mix with the groundwater. But in the last decade of work on shale gas in the United States, such cases have not been reported, except for the presence of methane gas in water sources. It is always better to monitor such events, and in case of any accident, immediate steps should be taken to stop the fracturing process or inform the community about such accident so that an alternate arrangement can be carried out till the problem is solved. To avoid such an accident, hydraulic fracturing should be avoided in the 5–8 m of the upper strata so that they can work as a cap rock and not allow the migration of either fluid or gases upward.

In all hydraulic fracturing, 40–60% water is with the shale formation, and the remaining comes back as flowback water initially with high pressure then later slows down. The flowback water will have many dissolved solids from the shale formation with very high pH and TDS. But with proper treatment, this water, along with some fresh water, can be used for hydraulic fracturing. There is large-scale air pollution from conventional and nonconventional hydrocarbon exploration. In the case of conventional hydrocarbon, the reservoir is in high gas pressure, and control of such gas escaping from the drilling well is monitored, and proper mechanisms are used to avoid escaping of gases to the environment. In the case of shale gas which is also under high pressure, there will be lot of volatile material that might escape along with the flowback water; to avoid such migration of volatile matter, some monitoring systems and devices separate the gases phase from the liquid phase in flowback water so that the gas phase can be controlled. Besides the emission from a shale gas well, there is continuous emission at the well pad by various compressors that run all the time during drilling; this can be controlled by using a proper trap mechanism so that emission of these gases to the environment can be reduce to a minimum. If large numbers of vehicles are used for the transport of water, which is also a source of air pollution, the effect can be reduced by using a pipeline instead of road transport, and it will also reduce noise pollution. Compared to conventional gas exploration, shale gas exploration has very limited emission to the environment.

The continuous use of power generators and air compressors is responsible for the emission of carbon and sulfur to the environment, and this has been classified under greenhouse gases. To reduce such emission, use of diesel can be reduced by using electrical power or solar power in the case of a very remote area. After the hydraulic fracturing, there is emission of methane in the atmosphere which is part of greenhouse gases. To avoid emission of methane into the environment; it can be trapped and sold as a source of energy. This will be good for our environment, and at the same time, industry can get money by selling to the user. It has been observed that in the case of conventional hydrocarbon these gases are burnt, which is again the emission of greenhouse gases to the environment. In fact, the methane from the well can be used to generate power at the site, which will reduce the consumption of diesel and indirectly the emission of greenhouse gases to the environment.

The exploration and exploitation of shale gas in the United States has reduced greenhouse emission to a large extent. The main reason is that coal as a fuel is being replaced by shale gas in major powerhouses in the United States, and China also plans to replace coal by shale gas to reduce emission levels in the metro areas. Producers like Arizona can also use shale gas in their powerhouse, and such an attempt will be a major contribution for a better global environment. Some environmentalists believe that emission of natural gas will be increasing with the growth of shale gas development and, indirectly, greenhouse emission will also increase.

6.9 INDUCED EARTHQUAKES

Induced seismicity is produced in a shale gas play area during a seismic survey and hydraulic fracturing. Though the intensity of seismicity is never than more than 4 on a seismic scale, even this seismicity is enough for groundwater contamination and release of volatile material from the shale gas play area. In the United States, due to a very low density population, no major impact to community has been reported. But countries with high density population will have severe problems with this seismicity. Though some reports have come for minor damage from seismicity in the United States, in the absence of base data, it is difficult to confirm that these events are due to shale gas exploration. During hydraulic fracturing, an increase of volume of the shale formation will indirectly produce seismicity. Similarly, disposal of flowback water after treatment and injection into abandoned deep wells also develops induced seismicity. Monitoring of this seismicity is necessary for a shale gas play area to avoid any measure of damage to property, human, and other living beings.

6.10 HAZARD MANAGEMENT

A seismic survey for subsurface study is responsible for induced seismicity and emission to environment. In shale gas exploration, initially, a small quantity of water is used for drilling a bore well and later a measured amount is used for hydraulic fracturing. For the transportation of water and heavy equipment, road infrastructure is used. In fracking fluid, chemicals are added, and flowback water is highly contaminated. These are the measured hazard in shale gas exploration. Induced seismicity is a major hazard during hydraulic fracturing compared to the seismic survey. During hydraulic fracturing, monitoring of induced seismicity is necessary, and a proper log book should be maintained by the operator. All the seismic stations should have networking in a control room so that the fracking area can be monitored from the control room for any accident. Monitoring is also necessary while injecting flowback water into deep wells. These are two major hazards for induced seismicity. Seismicity and their impact are given in the following table:

Magnitude	Impact from the Seismic Event
<2.5	Usually not felt, but recorded by seismograph
2.5–5.0	It is felt, but no damage is reported
5.0–6.0	It is felt, and there is damage to the structure
6.0–7.0	Felt strongly and may be minor damage to the structure
7.0–8.0	Felt strongly and major damage to the building and road
>8.0	Earthquake at a large scale can produce a tsunami and severe damage to infrastructure

The next hazard is contamination of shallow water aquifers by fracking fluid while hydraulic fracturing. During hydraulic fracturing the injection of fracturing fluid at high pressure activates small fractures, and many cracks in the upper rock will allow the fracking fluid to migrate upward, and later this fluid may contaminate the groundwater being used by the community. This is a major hazard, and a good amount of precaution and monitoring is necessary. If some small event is reported, immediate action is must, and that is possible only when base data of the shale gas play area is available and there is proper monitoring of all the shallow water aquifers in the operation area. For the safety of the human/living beings and aquatic animals, protection of shallow water aquifers is necessary to avoid any major health hazard. It is necessary that chemicals added in the fracking water as additives should not be hazardous to the community.

Management of flowback water and proper treatment is a major part of the shale gas exploration as this water is major hazard to the community. Proper monitoring and guidelines for treatment are required, and disposal of this water to surface water should be permitted only when this water has been cleaned as per the required guidelines. Water is the largest used substance, and very strong guidelines are required to protect it for any contamination by any means. There should be a specific tax to the industry using water in their processing that includes power generation, chemical industry, hydrocarbon industry (which includes shale gas exploration), food industry, paper industry, alcohol industry, etc. To protect water from major hazards, it is necessary for the industry to have a water treatment plant, and they can dispose the water from their industry only after the required treatment as per the guidelines issued by the civic authority.

One of the important parts of any process industry is the hazards to the people associated with the various processes. It is necessary to provide them prior training for the work so they can monitor it, and in case of emergency, they can take appropriate action. Persons working in handling flowback water should know about the chemicals being added and their impact to living beings. Associated people should take all required precaution while handling fracturing fluid and flowback water. Flowback water comes with high pressures with lots of semisolids as well as volatile material. Operation of flowback water needs trained people who are well aware about the associated hazards. In a shale gas operation, this is a major hazard, and well-trained personal are required for this operation.

As the hydraulic fracturing work is carried out with high air pressure [18,000–20,000 per square inch (psi)] the problem related with high pressure could be hazardous in case of any accident. For shale gas exploration and exploitation, training of the workers for various processes is most important for the sake of various hazards. As the system is associated with flammable gases and air at high pressure, various volatiles released after fracking need specific handling. To avoid any major accident, all the major high pressure lines should be designed with a bypass system, so in case of any accident, the bypass system can be used, and if possible under the monitoring system, this bypass should be a self-operating system. Inspection of all the necessary checkpoints is necessary for starting any process. It is also important to follow the regular maintenance schedule for all the equipment and other process systems. Such precautions will prevent any major accidents during the operation. There should be an arrangement for fire extinguishers as there is emission of flammable and volatile material that is hazardous. Storage of diesel and other flammable material should be at a specific location with a clearly marked boundary and proper security. All the shale gas play areas should be marked with a big display board for various cautions for the general community in the region. The area should be restricted for any flammable materials like match boxes, smoking, any acid battery, mobile phones, and other devices.

To avoid any gas leakage, a safety system should be fixed at all the necessary points at the well pad. In case of any excess gas or volatile materials, the alarm system should be actived to alert the workers. For excess pressure, there should be a provision for an alarm, and it should inform the operator for taking necessary action. For all the equipment, there should be a display board for starting and shutdown procedures for safety reasons. There should be a dress code for various operators including flame-resistant clothing for the people working near a fire hazard area.

6.11 WASTE WATER DISPOSAL

Disposal of flowback water after the necessary treatment is most important challenge in well completion. As the water is stored in a large quantity, the chances of spilling are also very high. In case of spillage of fracking water the harmful chemicals may seep into the groundwater. Violation of any safety norms is bound to create problems related to that process. Though, now, the industries are trying to use chemicals that are friendly to living beings. To reduce the quantity of fresh water, many industries are using recycling of flowback water after necessary treatment. But there is a limitation about using recycled water. Sometimes, the TDS of flowback water is very high, and it is difficult to use after the necessary treatment. Economy plays an important role for using treated water.

Besides using the flowback water, another option is to inject into deep wells or brine wells. But for the disposal of water in a deep well, there are guidelines, and under the guidelines, permission is required from the civic authority. But this option may not be available at all locations in the basin. It also depends on the geomorphology of the shale play area. In a structurally disturbed area and highly fractured area the injected water in a deep well may migrate upward and may contaminate the shallow water aquifers. The flowback water can be treated and reused for agriculture purposes and infrastructure projects. If the TDS is very high, mixing with fresh water is necessary to reduce the TDS per the disposal guidelines issued from the civic authority.

Shale gas exploration and exploitation is a large water-consuming industry. For one well, roughly 2–3 million gallons of water is required. Though this figure is large, compared to the daily water consumption in any metro area, which is between 500 and 800 million gallons of water, the quantity is not all that large. But in a drought-prone area, this quantity is very large, and industry has to look to minimize the water consumption. Regarding the chemicals used by the industry, now, it is mandatory to give the name of the chemicals and quantity for each fracking under the latest guidelines. It is no longer a trade secret for the industry. The industry is also supposed to give information about the various effects related to health in case of exposure.

It is necessary to take care with the high-pressure pipe while passing through shallow water aquifers. Use of proper casing and testing of casing prior to use is under the industry guidelines. Normally, water testing is carried out for at least 25% extra water pressure to avoid any leakage to shallow water aquifers. Presence of methane gas in the water pipeline for the nearby community is very common observation.

Case Studies

A.M. Dayal

CSIR-NGRI, Hyderabad, India

CONTENTS

Shale Gas. http://dx.doi.org/10.1016/B978-0-12-809573-7.00007-X

Producing Countries

7.1.1 UNITED STATES OF AMERICA

7.1.1.1 Introduction

For the first time in world history, shale gas was extracted in Fredonia, New York, United States of America (USA), in 1825. In 1915, Devonian shales from Big Sandy field produced shale gas at shallow depth in Floyd County, Kentucky. In 1965, hydraulic fracturing was started by a few operators at relatively low pressure, but they were disappointed as the production was low. In 1976, 5000 wells were drilled in south Virginia for a shale gas play in Ohio and Cleveland shale. There were shale gas plays in North Texas in 1981, but the production was not economical. The first commercial production from this region was in 1998.

7.1.1.2 Development of Shale Gas

For any industry, it is important to know the relationship with investment and income. For shale gas as well as any petroleum industry, first we will assess the area using various geophysical exploration tools like shallow seismic to know the thickness of shale in the basin, variation in shale formation, and depth of the shale. The next step is drilling of exploratory wells to obtain core samples for various geological, geochemical, and geomechanical studies. These studies will be useful to understand the physical and chemical properties of the shale formation, which will help in assessing the resource potential. Geochemical data will provide information about the total organic content, type of kerogen, and thermal maturity; major element data will provide information about the total silica and alumina content; petrophysics will provide information about permeability. Brittleness, porosity, and mineralogy will be useful to know the presence of different minerals in the shale formation.

The geochemical, geological, and geophysical information are useful parameters to assess the basin for commercial activity. If the data is promising, the operator looks for the location of a well pad. The well pad area should be big enough to install heavy equipment like generators, compressors, drilling rigs, reservoirs for various type of water storage, a water treatment plant to treat the flowback water, and storage of chemicals. This activity also requires infrastructure development like construction of roads for the transportation of equipment as well as the water from the source area. Construction of water reservoirs need a proper lining to avoid any contamination to ground water.

From one vertical well, an operator can drill multiple horizontal drilling up to a few kilometers and cover an area of 60–80 km². The combination of vertical wells and horizontal wells makes the shale gas play more attractive because from one well pad, hydraulic fracturing can be carried out in much bigger area, and production of gas will compensate the initial exploration cost. Hydraulic fracturing needs a large amount of water, sand as proppants, and some chemicals that will help smooth the hydraulic fracturing process. The basic concept of hydraulic fracturing is to open the individual shale layers so that trapped gas can be exploited. Fracturing the shale formation at depth needs hydraulic fracturing at high pressure (18,000–20,000 psi). The water is injected into the shale formation using high-capacity pumps. Composition of fracking fluid is mainly water with sand and chemicals. Chemicals are added to increase viscosity and offer protection to the pipes from corrosion. Sand as a proppant is required to prevent the closing of fractured shale. For the safety of shallow water aquifers the hydraulic fracturing is allowed at a depth of more than 1000 m. This will not allow the contamination of shallow water aquifers. Once the well goes into production mode, the normal life of the production well is 15–20 years. During the production, all the wells are monitored for any contamination to ground water and leakage to the environment. Once the production from the well is over, the well is cemented for safety. The well cores are disposed of at a proper place, and all the flowback water is treated before surface disposal. The play area is cleaned and handed over to the owner for the agriculture or another purpose.

Faced with declining natural gas production, the federal government financially supported the development of an alternative supply of energy, which also included the Eastern Gas Shales Project from 1976 to 1992. For the research and development, additional funding was allocated to the Gas Research Institute for one decade. The Department of Energy in partnership with industries drilled the first successful horizontal well in the Barnett shale formation in 1986. With the initial success in Texas, many industries were attracted toward the exploration and exploitation of shale gas in the USA. As the shale gas is environmentally friendly, there was strong support from the government also. With the current rate of production of shale gas in the USA as well as a few other countries, it will last for 100–250 years.

Shale gas production is the USA has grown rapidly in the last decade with an excellent partnership of industry and researchers. They developed the horizontal drilling and hydraulic fracturing at very high pressure to frack the shale. From 1996 onward the production of shale gas increased slowly, and after 2006 the production of shale gas was very rapid, which is 5.9% of US gas production. From 2006 onward, there was large-scale drilling of wells in different basins in the USA. The basic reason was the development of expertise by petroleum industry for horizontal drilling and hydraulic fracturing. This allowed them to produce shale gas in large amounts, and large-scale production of shale gas was responsible for the downfall of gas prices from $10 per million BTU in 2010 to $2.5 per million BTU in 2014.

In 2008 the production of shale gas in the USA was 2.02 Tcf per year, which is a growth of 71% over the past year. In 2009, the USA produced 3.11 Tcf per year of shale gas, which is again a growth of 71% over the year 2008. At the same time, continuous efforts for the shale gas exploration in 2009 increased the production to 76% (60.6 Tcf). In 2015 the production of shale gas was 41.1 Bcf per day. From March 2015 onward, production of shale gas has had an increasing trend (Fig. 7.1.1).

Research and development is the key of innovation. Shale gas exploration and exploitation in the USA from 2007 to 2014 was an excellent growth of the industry. With development of expertise the industry took a number of shale formations in different basins in the USA and increased the production of shale gas

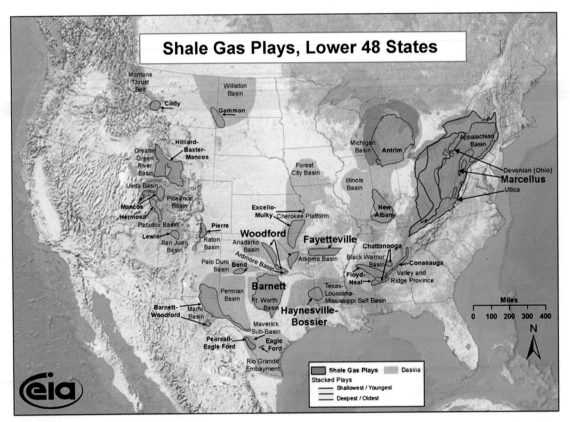

FIGURE 7.1.1
Shale gas wells in the United States of America.

beyond imagination. Along with the gas, they also concentrated on shale oil, and as the gas was available, shale oil production also increased remarkably.

7.1.1.3 Shale Formations Under Production in the United States

The Upper Devonian Antrim shale is in the Michigan basin. The first shale gas was produced from Antrim shale in the 1940s, the play was of economic importance till the late 1980s. The natural gas from Antrim shale is of biogenic origin generated by the action of bacteria on an organic-rich shale formation. With the development of new expertise on shale gas exploration, these shales were exploited using new technology from 2007 onward, and it is the 13th largest shale gas play in the USA.

Barnett Shale, Texas

The first well for shale gas was drilled in 1981. With increasing prices of natural gas, a new area was taken for exploration and exploitation. The new technology for shale gas was responsible for the large-scale production. The average thickness of the Barnett shale is 300 ft, though at some places it is almost 1000 ft. Based on the present work, it has been observed that shale formations between 300 and 600 ft deep with a thickness of 180 ft are the most economic wells. In 2007, a gas field in the eastern USA produced 1.11 Tcf of gas, which was second largest and 6% of US natural gas production.

Caney Shale, Oklahoma

The Caney Shale, which is equivalent to Barnett shale in the Arkoma Basin, is producing gas for commercial purpose. This shale gas play is a larger success than the Barnett play.

Conesauga Shale, Alabama

In Conesauga shale, Alabama, wells were drilled during 2008–2009 for shale gas exploration and exploitation. The shale formation in the basin is producing shale gas for commercial purpose. Fayetteville Shale in Arkoma basin in Mississippi is 50–550 ft thick with a depth that varies from 1500 to 6500 ft. The shale gas was produced earlier in vertical wells, but now with drilling of horizontal wells in Fayetteville, shale gas is under production.

Floyd Shale, Alabama

The exploration and exploitation work is going on to produce shale gas at a commercial level from the Floyd Shale.

Gothic Shale, Colorado

USGS in 1916 reported the oil-bearing shale from Gothic shale in Colorado. They reported 20,000 million barrels of crude oil. These shales were explored and exploited for shale gas production. Presently a horizontal well in the Gothic shale is producing 5700 Mcf of gas per day.

Haynesville Shale, Louisiana

The production of shale gas from Haynesville shale in northwest Louisiana was reported in 1905. Further work was carried out in 2007 onward from vertical and horizontal drilling in the northeast of Texas for Bossier shale, which is equivalent to the Haynesville shale.

Collingwood Utica Shale, Michigan

Collingwood Utica shale, Michigan, has a very good reserve for shale gas play. This is the most promising oil and gas play in the USA.

Albany Shale, Kentucky

The Albany shale in Kentucky is a 100-year-old gas field. With the increasing demand of gas, these shales were exploited with hydraulic fracturing and horizontal drilling, and the drilling activity was increased in the basin area. The shale wells are 250–2000 ft deep, and the gas in the wells is of biogenic and thermogenic origin.

Pearsall Shale, Texas

In the Pearsall Shale of the Maverick Basin, south of Texas, 50 wells were drilled initially to study the resources for shale gas play. The gas in under production from this basin.

Albany Shale

Albany Shale is producing shale gas in the southern part of the USA. The gas play in Pennsylvania has more than 20,000 wells for exploration and exploitation, and these wells are producing gas at a commercial scale. The wells are 3000–5000 ft deep. The Chattanooga Shale, also called the Ohio Shale, is producing gas at a large scale. According to USGS, total reserves in Devonian black shales is 12.2 Tcf from Kentucky to New York.

Marcellus Shale

The estimated reserve of gas in the Marcellus shale in West Virginia, Pennsylvania, and New York is 168–516 Tcf. With a large shale gas reserve in the basin, it is estimated that it will be able to supply gas requirements in the northeast part of the USA.

Woodform Shale, Oklahoma

Earlier, there was gas production with conventional technique, but by 2008, more than 750 vertical gas wells were drilled. There are number of operators for shale gas plays, and in 2011 the production of shale gas was at its peak.

In 2011, according to US Energy Information Administration (EIA), the recoverable reserves in the USA increased to 827 Tcf, which includes the new shale data from Marcellus, Haynesville, and Eagle Ford shales. The US government has allowed them to convert the coal-based energy plants to gas-based energy plants to reduce carbon dioxide and sulfur emission.

The USA can supply shale gas to Europe for the long term with LNG at $7–8 per million BTU based on long-range marginal costs for additional infrastructure (US LNG Marketing's vice president for strategy, May 25, 2016). He feels that this is a very effective negotiating offer for the European Union compared to its existing suppliers like Russia and Norway. LNG supply in the USA is already able to compete with Europe's existing gas supply sources at current prices of $4–4.50 per million BTU based on the current infrastructure capacity. The USA is exporting shale gas as LNG to South America, the Middle East, and Asia.

The first US shipment of natural gas to the European Union has been received at the Portuguese port of Sines (Galp Energia). Galp said it had bought enough gas to be good for a week of Portuguese consumption or 2% of annual demand. With the development of shale gas, the United States has become the world's biggest gas producer and is set to become a net exporter this year. US gas has already being shipped to Argentina and Brazil, as well as India.

PRODUCING COUNTRIES

7.1.2 CANADA

There are large reserves of conventional gas in Canada, and with the inclusion of shale gas reserves (573 Tcf), the total estimated recoverable reserves of gas in Canada have gone up (Fig. 7.1.2). In 1858, more than 30 wells were drilled in Oil Springs, Ontario, for commercial oil production. Since then, a number of wells were drilled for oil and natural gas in Canada. First, oil from shale was discovered on 1921. With the increasing demand and new technology, the first shale gas play was started in the region of British Columbia in 2005. With a large demand for oil and gas, there was an increase in the production of shale gas and tight oil in the western part of Canada (Fig. 7.1.2). The major shale gas reserve are given next.

Utica Shale, Quebec

The Utica Shale of Ordovician age in Quebec has reserves of 4000 Bcf with a production of 1 Mcf per day. New 24 vertical and horizontal wells were drilled between 2006 and 2009 to test the Utica shale. The analytical data indicated positive results, though till 2009 there was no production. The Utica shale is a black calcareous shale with a thickness of 150–700 ft, high total organic carbon (TOC) content, and gas reserve of 31 Tcf.

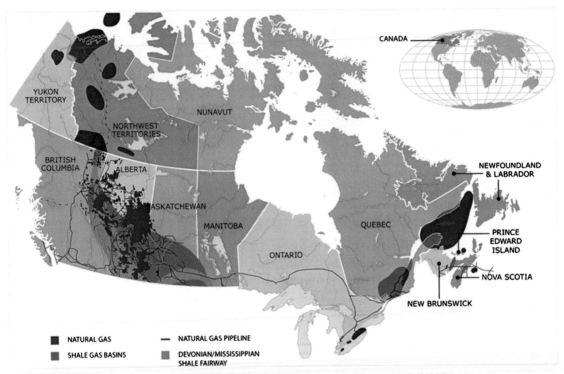

FIGURE 7.1.2
Shale gas play in Canada.

Muskwa Formation, British Columbia

The shale in https://en.wikipedia.org/wiki/Muskwa_Formation Muskwa for mation of Devonian age in northeast https://en.wikipedia.org/wiki/British_ Columbia Canada has estimated reserves of 6000 Bcf of recoverable gas.

Duvernay Formation, Alberta

The shale in Duvernay formation is the source rock for light oil in Alberta. Shale gas and condensate are produced using horizontal drilling and multi-stage hydraulic fracturing.

New Brunswick Oil and Gas Field

This is one of the oldest oil fields in Canada with a commercial production in 1909. Now with the shale gas development in the USA, exploration and exploitation of shale gas is under progress.

Frederick Brook Shale

New Brunswick has an estimated reserve of 80 Tcf of gas with a depth of 1000 m. This formation age is 300 Ma. The exploration for shale gas is a joint venture with Corridor Resources Inc., who is drilling wells for gas. Two vertical wells were drilled.

The production of shale gas in Canada is 4% of their demand. There is good reserve of tight gas that constitutes 47%, and in future, it will represent 90% of Canada's natural gas production.

PRODUCING BASINS

7.1.3 CHINA

China has the largest shale gas reserves (1115 Tcf) in the world (Fig. 7.1.3). So far, shale gas productivity exceeds 6 billion cubic meters (Bcm) in the four major producing areas, including Fuling, Changning, Weiyuan, and Yanchang.

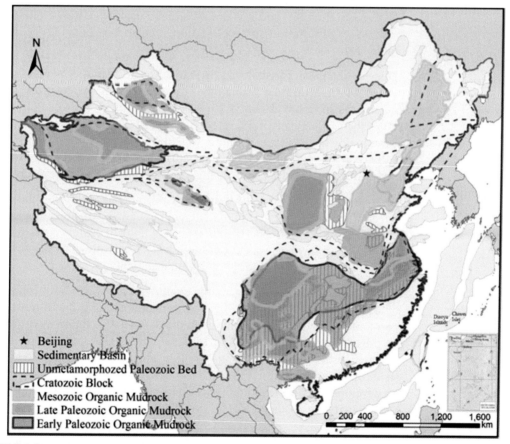

FIGURE 7.1.3
The distribution of organic rich shale gas reservoir in China.

China's shale gas output was 25 million cubic meters (Mcm) in 2012 and increased sixfold in 2014 to reach 1.3 Bcm. Beginning in 2015, shale gas exploration in China went up, and the output is expected to top 5 Bcm this year, and China will become the third country in the world for shale gas commercial development after the USA and Canada.

Currently, the exploration area of shale gas in China covers 170,000 km^2, mainly in the Sichuan basin and its surrounding areas. For the exploration and exploitation of shale gas in Sichuan basin, 840 new wells were drilled. In China, all the basins are in rugged terrain, and it is difficult to transport drilling rigs and other material; that is the reason that one of the largest reserves of shale gas has been unable to be exploited.

Still, expertise in drilling technology is in the USA, and under a joint program with various US petroleum companies, Chinese oil and gas companies are drilling wells for shale gas exploration. The shale formations in China are not like those in the USA, and it is not possible to use the existing US technology for shale gas exploration and exploitation. Sinopec has a target of 6.5 Bcm per year by 2016.

Achieving the target for shale gas by Petrochina and Sinopec would be a great success for the country as China plans to use these energy resources to reduce emissions. China has declared 2016 as the year of a "War on Pollution" to reduce the various emission levels for carbon dioxide, sulfur, and nitric oxide in the major metro areas of China. By 2017, China is aiming to provide 9% of the total energy demand through natural gas. Toward this goal, China has already made some progress by using a greater quantity of natural gas.

The shale gas industry is in its initial stage. Compared to the USA the exploration and exploitation costs are much higher in China, which includes drilling. But with gaining experience, China has brought down drilling costs quite a lot. In 2012 the drilling cost was $14 million, which came down to $11–13 million in 2015. Operating cost comes down with experience.

In 2015, China was importing a large quantity of gas worth $400 billion from Russia ($10 per million BTU). This import may slow down the in-house exploration and exploitation of shale gas. With excess gas in the market the global gas prices have come down $5 to $8 per million BTU. But if China develops this large resource of shale gas, they will be able to control the prices of shale gas in the international market. China is spending a large sum of money for the development and exploitation of shale gas, but fluctuating gas prices and water scarcity are obstacles.

From 2014 to 2015 China had increased its recoverable shale gas reserves more than fivefold. The total recoverable shale gas reserves is 130 Bcm. They have plans to replace coal and conventional oil-based powerhouses with shale gas by 2015 onward. In 2015, against the target of 6.5 Bcm, the production was 4.47 Bcm, which is more than a threefold increase. British Petroleum signed its first shale gas production deal, joining with China's oil company. For 2016 the target is to produce 200 million metric tons of crude oil and 144 Bcm of natural gas. For conventional and nonconventional energy resources, the production in 2015 is given next:

- In 2015, the recoverable crude reserve was 217 million tons.
- Total recoverable conventional natural gas reserves are 5.194 Tcm. The output in 2015 for conventional natural gas was 124.5 Bcm.
- China has 306 Bcm of CBM reserves.
- According to Sinopec on May 25, 2016, their target is to triple the shale gas capacity in Chongqing in southwest China to 15 Bcm per year by 2020 from the current 5 Bcm per year. It also aims to increase its shale gas output in Chongqing to 10 Bcm per year by 2020.
- China Sinopec has completed a shale gas target of the 5 Bcm per year in Phase I of its flagship Fuling project in Chongqing in 2015 and is working on the second phase in the year 2016.
- The proven developed reserves in Fuling are 28.77 Bcm, which is more than double its earlier reserve of 13.37 Bcm.

Chapter 7.2

Prospective Countries

7.2.1 ASIA

7.2.1.1 India

India has 28 sedimentary basin and an estimated reserve of shale gas of 500–625 Tcf (Fig. 7.2.1). In 2016, there is a change in policy, and in the future, all the blocks given for exploration can be explored for conventional and nonconventional energy resources, which includes conventional and nonconventional hydrocarbon, coal bed methene, and coal itself. There was meeting of the Directorate General Hydrocarbons with USGS to identify the shale gas resources for the exploration and exploitation of shale gas in 2010. The USA has agreed to work together for the development of clean energy technologies.

State run Oil & Natural Gas Corp (ONGC) has produced the first shale gas in the country at Icchapur in Durgapur district of West Bengal. According to ONGC the 2000-m-deep well at the shale formation encountered gas on

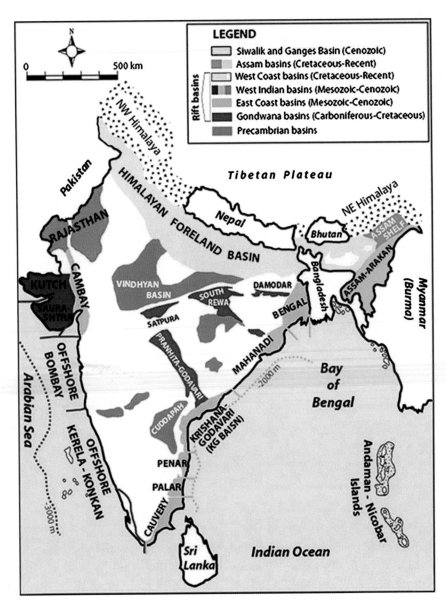

FIGURE 7.2.1
Sedimentary basins in India.

January 25, 2011. On August 10, 2012, the government released its draft policy for the auction of blocks of shale gas and oil. India has several shale formations that could hold gas. As per the draft, shale gas could be present extensively in sedimentary basins such as Cambay, Gondwana, Krishna-Godavari, Cauvery, Cuddapah, and Vindhyan. Shale gas production in India will boost the country's economy. India too seems to be getting its act together with the petroleum ministry finalizing the shale oil exploration policy, which will be approved by the cabinet. DGH is likely to auction 72 marginal blocks in 2016 for the exploration of conventional and nonconventional energy. This will be the first auction for blocks for shale gas to private players.

The prime minister stressed taking a fresh look at the petroleum sector during a meeting with global oil and gas experts on January 5, 2016. There was a meeting of the Indian prime minister with the chief of Bharat Petroleum and top executives of Shell International Energy Agency members to discuss ways of boosting investments during times of low oil prices. The interaction, which lasted over 2 h, also included union ministers for finance, energy, and petroleum, the vice chairman of NITI Aayog, and a few other experts. The prime minister emphasized his vision for a fresh look at the sector to bring investment, technological upgrade, and development of human resources.

India is hoping to unlock its shale gas reserves by inviting investments from private companies on a profit sharing basis. The country is believed to have more than 625 Tcf of shale gas reserves. These gas reserves are enough to run the country's gas-fired power stations for the next few decades. But it may take 5–7 years for the country to access and realize profits from this valuable natural resource because of a lack of infrastructure, opposition to raising gas prices, and lack of information about exactly where to find the gas.

CASE STUDY FOR CAMBAY BASIN, INDIA

U. Vadapalli and N. Vedanti

CSIR-NGRI, Hyderabad, India

7.2.2 CASE STUDY: WELL LOG ANALYSIS FOR SHALE GAS PROSPECTS IN ANKLESHWAR FIELD, CAMBAY BASIN, INDIA

7.2.2.1 Introduction

Successful exploration and exploitation of shale gas from Barnett shale formation in the United States of America (USA) has created considerable excitement globally to find the next Barnett shale. Shale gas exploration in the USA has become a game changer for making the country self-sufficient in natural gas over the last few years. To know the world's shale gas resource potential, Advanced Resources International, Inc. (ARI) for the US DOE's EIA has conducted the World Shale Gas and Shale Oil Resource Assessment (ARI and US EIA, 2011; Kuuskraa et al., 2013). The study has identified 48 major shale basins in 32 countries, which include Canada, Mexico, North and South

American countries, the United Kingdom (UK), China, India, Pakistan, and Australia. In India, the Cambay basin, Krishna-Godavari basin, Cauvery basin, and Damodar valley are the priority for shale gas exploration (Kuuskraa et al., 2013). India has an obvious interest to explore shale gas domestically to fill in the gap between demand and supply of energy. Oil and gas companies of India viz., Reliance Industries Limited (RIL), Oil India Limited (OIL), Indian Oil Corporation (IOC), and Gas Authority of India Ltd. (GAIL) have invested in shale gas production in the USA. In 2012 the government of India made a policy for exploration and exploitation of shale gas in India (Batra, 2013). Recently, ONGC Ltd., India also ventured into shale gas exploration. The first exploratory well of India was drilled by ONGC in Ankleshwar asset of Cambay basin on October 27, 2013 (ongmazdoorsangh.org). The Cambay shale formation is the source rock for oil and gas deposits discovered in Cambay basin. According to ARI and US EIA (2013) the Risked Gas In-Place in Cambay shale formation is 146 Tcf, of which 30 Tcf is technically recoverable. The Ankleshwar oil field of Cambay basin is one of the mature and highly producing oil fields. The location map of Cambay basin along with its oil and gas fields is shown in Fig. 7.2.2.

ONGC found that some of the wells were logged beyond the base of the Cambay shale formation, and here, out of interest, these shales were analyzed further for shale gas. The Cambay shale formation has already been proven to have shale gas resource potential (ARI and US EIA, 2011; Biswas et al., 2013; Kuuskraa et al., 2013; Padhy and Das, 2013; Sain et al., 2014). In these wells the Cambay shale formation has been identified using gamma ray, density, and sonic logs, and these logs were analyzed to identify the occurrence of shale gas. ONGC found indications for occurrence of shale gas in three wells.

7.2.2.2 Shale Gas Reservoir Characterization

The organic rich shales, which have undergone through 160–225°C, are good for shale gas. Shale gas reservoir characterization involves geochemical, petrophysical, geomechanical, and geophysical analysis. The properties need to be evaluated for the identification of shales with production potential and thermal maturity. The relationship with organic carbon content and resource potential is given in Table 7.2.1 A high level of TOC is a critical factor to assess shale gas reservoirs. Petrophysical analysis used for exploration of shale gas involves the use of well logs and borehole image logs. Shales with hydrocarbon production potential display specific characteristics on well logs, which can be used to discriminate prospective shales from shales with little or no potential. Geomechanical analysis helps to identify them. Geophysical studies help to identify the areas of shale gas prospect potential and sweet spots of TOC beyond the well locations.

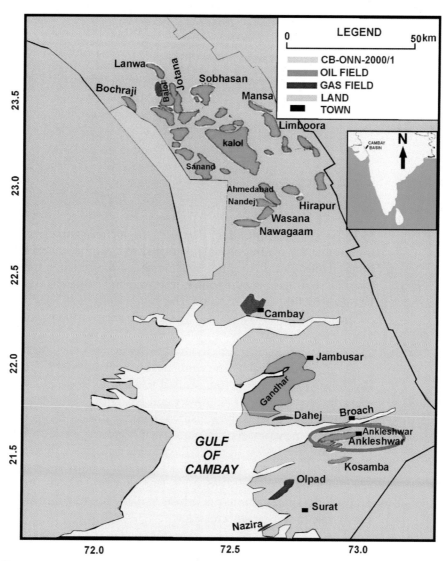

FIGURE 7.2.2

Location map of Cambay basin showing its oil and gas fields. Ankleshwar field is highlighted by *ellipse*. Ankleshwar field trending ENE-WSW is highlighted in the figure. *Adapted from* www.spgindia.org, *2013.*

7.2.2.3 Well Log Indications of Shale Gas

The basic well logs required to give primary information about the shale gas potential in a well are gamma ray, resistivity, sonic, bulk density, and neutron porosity. The gamma ray reading is the combined response of potassium, thorium, and uranium in the formation. Since shales have a higher concentration of potassium, they can be distinguished from the conventional reservoir

Table 7.2.1 The Relationship Between the Total Organic Carbon (TOC) and Resource Potential (Alexander et al., 2011)

TOC (Weight%)	Resource Potential
<0.5	Very poor
0.5–1	Poor
1–2	Fair
2–4	Good
4–10	Very good
>10	Unknown

rocks (sandstone, limestone) based on the higher gamma ray reading. Thus, the gamma ray reading in organic-rich shales can be higher than in conventional shales. But this cannot be generalized for all shale formations. In the case of a shale gas play the source, reservoir, and seal are contained within the fine-grained lithofacies of shale. The gamma ray curve may or may not discriminate the three. The resistivity of a formation is majorly affected by fluids inside the pore space. Since organic matter has high resistivity, the organic-rich shales possess higher resistivity than the conventional shales, which contain clay-bound saline water. The sonic transit time–resistivity crossplots or density–resistivity crossplots can be used to identify organic-rich mature source rocks and to calculate TOC (Meyer and Nederlof, 1984; Passey et al., 1990). Porosity measurements show distinct character in gas-bearing shales. The presence of gas increases density porosity and reduces neutron porosity due its lower hydrogen index. In general, conventional shales, due to higher density and higher neutron porosity, exhibit a uniform positive separation on the density and neutron porosity crossplot. Due to lower density and lower neutron porosity, the shale gas intervals exhibit variable negative separation on the density and neutron porosity crossplot.

7.2.2.4 The Study Area

The study area of Ankleshwar oil field is one of the major oil producing fields of Cambay basin, India (Fig. 7.2.2). The oil field, located 16 km southwest of Broach town in the alluvial plains of the Narmada River, was discovered by ONGC in 1959. In the field, tertiary sediments varying in thickness between a depth of 1343 and 2026 m have been deposited over Deccan trap basement (Mukherjee, 1981). The stratigraphy of the Cambay basin is shown in Fig. 7.2.3. The oil-bearing Cambay basin is a graben structure located in the western Indian state of Gujarat (Fig. 7.2.2). It lies between 21°00′ and 24°00′ north and 71°30′ and 73°30′ east. The graben is 5–7 km deep and has a width of 40–80 km (Chowdhary, 2004). It is a linear NNW-SSE trending rift, which is about 425 km

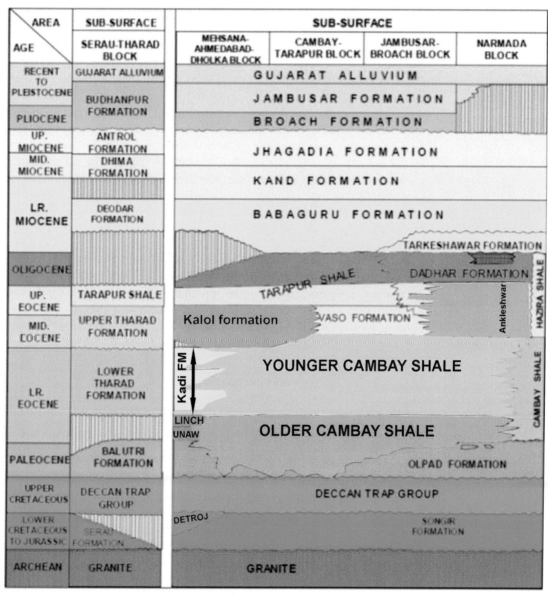

FIGURE 7.2.3

Generalized stratigraphy of Cambay basin showing stratigraphy of the Ankleshwar formation (www.dghindia.org).

long. The basin, including its flanks, covers an approximate area of almost 53,500 sq. km. Orogenic basement trends play an important role in the tectonic origin of Cambay graben (Biswas, 1987; Biswas et al., 1993; Chowdhary, 2004; Nanawati et al., 1995; Raju, 1968; Roy, 1990). These basement trends divide the basin into five tectonic blocks (Holloway et al., 2007; Nanawati et al., 1995;

Biswas et al., 1993): (1) Patan Tharad Sanchor block; (2) Mehsana-Ahmedabad block; (3) Tarapur block; (4) Jambusar-Broach block, and (5) Narmada block. In Cambay basin, Cambay shale formation is the regional source rock for oil and gas deposits. The thickness of the Cambay shale formation varies between 500 and 1500 m (Chowdhary, 2004). The Cambay shale formation consists of two subunits: (1) upper Cambay shale and (2) lower Cambay shale. These are separated by an erosional unconformity, known as a neck marker on electro logs (Pandey and Dave, 1998). The oil accumulation in the Hazad member of Ankleshwar reservoir is mainly generated and expelled from the source rocks in the upper part of Cambay shale (Chowdhary, 2004). The oil generated by the lower part of the Cambay shale may have already cracked into gas (Yalcin et al., 1988). The TOC content and thermal maturation suggest good prospects of shale gas in the Cambay shale formation of the Cambay basin (Padhy and Das, 2013; Dayal et al., 2013).

7.2.2.5 Shale Gas Prospect Identification in the Study Area

The Cambay shale formation in the study area has been proven to have shale gas reservoir properties and resource potential (ARI and US EIA, 2011; Sain et al., 2014). To identify the shale gas prospects in the study area, the Cambay shale formation has been identified using a gamma ray log and analyzed for signatures of shale gas on density, resistivity, and neutron porosity logs. The TOC within the Cambay shale interval is estimated from a density log using the widely accepted Schmoker's formula (Schmoker, 1980), which is given as follows:

$$TOC = \frac{154.497}{\rho_b} - 57.261$$

Three wells viz. W_1, W_2 and W_3 were drilled through Cambay shale formation and are used for the analysis. The Cambay shale interval in these three wells identified on gamma ray log is shown in Fig. 7.2.4A–C. Along with the gamma ray log, resistivity, bulk density (RHOB), density neutron crossplot, total porosity derived from density log, and TOC log are also shown in the figures. In Fig. 7.2.4A–C high TOC intervals are correlated with high resistivity, high total porosity, negative and variable separation on density neutron crossplot, and lower bulk density. In all the three wells the bulk density log has an increasing trend until the top of the shale gas prospect zone; beyond that it shows a decreasing trend. A sonic log is available for well W_2. The sonic transit time curve of W_2 has a decreasing trend until 1365 m; beyond this point, it shows a slight increment in sonic transit time. The trend change on density and sonic logs may indicate an overpressured condition in the Cambay shale formation, which is already documented in the literature (ARI and US EIA, 2011; Sain et al., 2014). In the case of well W_3 the ringing observed on, density, TOC, and total porosity logs (Fig. 7.2.4C) could be due to bad borehole condition, which is observed on caliper log.

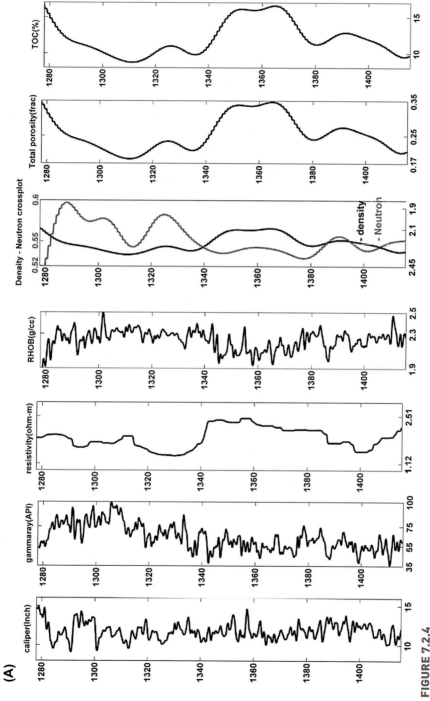

FIGURE 7.2.4

Figure illustrates the well logs response in the Cambay shale interval of wells (A) W1, (B) W2, (C) W3. Correlation among high resistivity, low bulk density, negative variable separation on density neutron crossplot, high total porosity, and high TOC can be seen. Sonic log also shows increased transit time in the zone of interest (B)

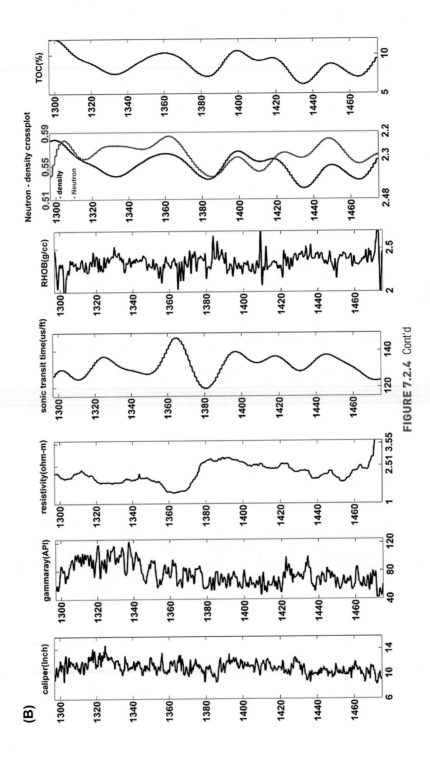

FIGURE 7.2.4 Cont'd

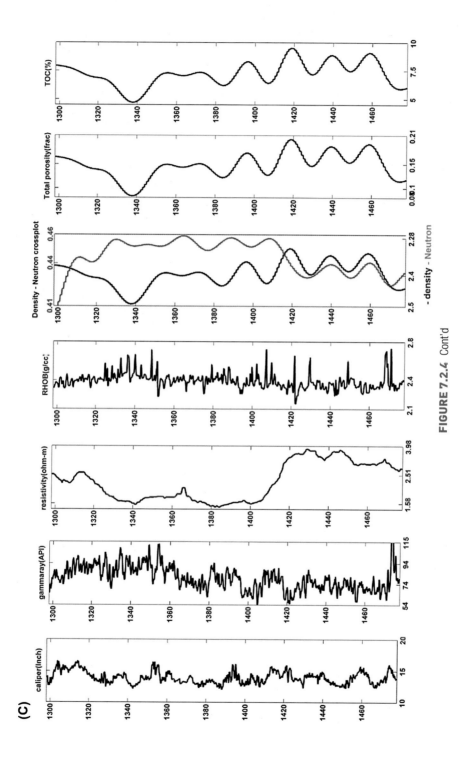

FIGURE 7.2.4 Cont'd

7.2.2.6 Procedure Followed for Log Analysis

The procedure of log analysis followed in the study is as follows:

1. identification of Cambay shale formation on gamma ray curve using high gamma ray reading,
2. identification of high TOC and high total porosity zones by estimating TOC and total porosity from density log. Before going to this step make sure that there are no caved zones indicated on the caliper log, which can affect the bulk density reading,
3. identification of high resistivity zone on resistivity curve,
4. identification of overpressure zone by using density log based on change in trend form increasing density to decreasing density,
5. identification of density neutron negative separation zone on density neutron crossplot.

7.2.2.7 Conclusions

Analysis of well logs from three wells drilled in Ankleshwar reservoir reveals that Cambay shale has a good prospect for shale gas exploration in the future. However, it is worth to mention here that production of gas from Cambay shale may not be an easy task as the formation may be very tight (Ramanathan et al., 2010). In such cases, hydraulic fracturing is required. Recently, many environmental organizations have raised concerns on the application of hydraulic fracturing for production of shale gas (Kumar and Shandilya, 2013). Hydraulic fracturing is a technique in which water mixed with sand and chemicals is injected at high pressure into a borehole to create smaller fractures (<1 mm) in the shale formation. This helps fluids trapped in shales to migrate toward the production well. Major environmental concerns raised by this technique include groundwater contamination and an increase in micro-seismic/seismic activity. Hence, a multidisciplinary approach is required for safe and economic production of shale gas.

References

Alexander, T., Baihly, J., Boyer, C., Clark, B., Waters, G., Jochen, V., Toelle, B.E., 2011. Gas shale revolution. Oilfield Rev. 23, 40–55.

Batra, R.K., 2013. Shale Gas in India: Look Before You Leap. http://www.teriin.org, 23.

Biswas, S.K., 1987. Regional tectonic framework, structure and evolution of the western marginal basins of India. Tectonophysics 135 (4), 307–327.

Biswas, S.K., Bhasin, A.L., Ram, J., 1993. Classification of Indian sedimentary basins in the framework of plate tectonics. In: Proc. Second Seminar on Petroliferous Basins of India, vol. 1. Indian Petroleum, Dehradun, India, pp. 1–43.

Biswas, S.K., Ariketi, R., Dubey, R., Chandra, S., 2013. Shale Gas Evaluation of Cambay Shale Formation in Tarapur Syncline, Cambay Basin, India – A Seismo-geological Approach. SPG Conference, Kochi, p. 142.

Chowdhary, L.R., 2004. Petroleum Geology of the Cambay Basin, Gujarat, India. Indian Petroleum Publishers.

Dayal, A.M., Mani, D., Mishra, S., Patil, D.J., 2013. Shale gas prospects of the Cambay basin, western India. Geohorizon 18 (1), 26–31.

Holloway, S., Garg, A., Kapshe, M., Pracha, A.S., Khan, S.R., Mahmood, M.A., Singh, T.N., Kirk, K.L., Applequist, L.R., Deshpande, A., Evans, D.J., Garg, Y., Vincent, C.J., Williams, J.D.O., 2007. A Regional Assessment of the Potential for CO_2 Storage in the Indian Subcontinent. Sustainable and Renewable Energy Programme Commissioned Report CR/07/198 by British Geological Survey (BGS). NERC.

Kumar, T., Shandilya, A., 2013. Tight Reservoirs: An Overview in Indian Context. SPG conference, Kochi, p. 260.

Kuuskraa, V.A., Stevens, S.H., Moodhe, K., 2013. EIA/ARI World Shale Gas and Shale Oil Resource Assessment. Advanced Resources International INC.

Meyer, B.L., Nederlof, M.H., 1984. Identification of source rocks on wireline logs by density/resistivity and sonic transit time/resistivity crossplots. AAPG Bull. 68 (2), 121–129.

Mukherjee, M.K., 1981. Evolution of Anklesvar anticline, Cambay basin, India. AAPG Bull. 65 (2), 336–343.

Nanawati, V., Jain, A.K., Jain, S., Singh, H., 1995. Efficacy of Olpad formation as source rock in Ahmedabad–Mehsana block of Cambay basin, India. In: Petrotech Conference. Technology Trends in Petroleum Industry by R B Publishing Corporation, New Delhi.

Padhy, P.K., Das, S.K., 2013. Shale oil and gas plays: Indian sedimentary basins. Geohorizons 18 (1), 20–25.

Pandey, J., Dave, A., 1998. Stratigraphy of Indian Petroliferous Basins. National Institute of Oceanography.

Passey, Q.R., Creancy, S., Kulla, J.B., Moretti, F.J., Stroud, J.D., 1990. A practical model for organic richness from porosity and resistivity logs. AAPG Bull. 74 (12), 1777–1794.

Ramanathan, V., Pankaj, P., Susanta, P.G.H., Dubey, S., Farooqui, M.Y., Chauhan, S.P.S., Singh, P., January 2010. Successful Hydrofracturing Campaign Leads to Commercial Development of Tarapur Field in Gujarat. In: SPE Oil and Gas India Conference and Exhibition. Society of Petroleum Engineers.

Raju, A.T.R., 1968. Geological evolution of Assam and Cambay tertiary basins of India. AAPG Bull. 52 (12), 2422–2437.

Roy, T.K., 1990. Structural styles in southern Cambay basin India and role of Narmada geofracture in the formation of giant hydrocarbon accumulation. Bull. Oil Nat. Gas Comm. 27, 15–38.

Sain, K., Rai, M., Sen, M.K., 2014. A review on shale gas prospect in Indian sedimentary basins. J. Indian Geophys. Union 18 (2), 183–194.

Schmoker, J.W., 1980. Organic content of Devonian shale in western Appalachian basin. AAPG Bull. 64 (12), 2156–2165.

United States Energy Information Administration, Kuuskraa, V., 2011. World Shale Gas Resources: An Initial Assessment of 14 Regions Outside the United States. US Department of Energy.

Yalcin, M.N., Welte, D.H., Misra, K.N., Mandal, S.K., Balan, K.C., Mehrotra, K.L., et al., 1988. 3-D Computer Aided Basin Modelling of Cambay Basin, India—a Case History of Hydrocarbon Generation. Petroleum Geochemistry and Exploration in Afro-asian Region. Rotterdam, Balkema, pp. 417–450.

CHAPTER 8

Environmental Concerns of Shale Gas Production

A.M. Dayal

CSIR-NGRI, Hyderabad, India

CONTENTS

8.1 INTRODUCTION

Carbonaceous shale is a fine-grained, organic-rich sedimentary rock. The technically recoverable gas depends on a large number of parameters like thermal maturity, mineralogy, amount of silica content, water source in a nearby area, and a facility for disposal of flowback water. Shale is also the source rock for conventional hydrocarbon. So far, we have been exploring and exploiting conventional oil and gas as they migrated from source rock to the reservoir rocks, and it was easy to explore them from such rocks. Geologist knew about the presence of free and adsorbed gas in organic-rich shale formations, but the exploitation technique was not available. With the increasing cost of oil and

Shale Gas. http://dx.doi.org/10.1016/B978-0-12-809573-7.00008-1

also declining sources of conventional hydrocarbon, it was necessary to look for an alternate source of energy. Solar, wind, and nuclear are alternate sources but cannot replace conventional oil and coal as a major source of energy. The United States took the lead to develop an alternate source of energy and set up the Shale Research Institute in the year 2000 to develop hydraulic fracturing and horizontal drilling. Initially, these techniques were used at a small scale, and with necessary improvement in 2008 onward the petroleum industry could take up the shale gas project on a large scale. This new source of energy has changed the total scenario of the world oil market. With this revolution the prices fell, and within four years the prices were 25% of the earlier prices. The large-scale production of shale gas in the United States has reduced the import of gas from Canada, and overall there is excess supply of oil and gas in the global market, which was responsible for the fall of oil prices.

8.2 INDUCED SEISMICITY

For shale gas exploration, three-dimensional seismic work is carried out to find the extent of shale formation and the thickness of different subsurface shale formations. To carry out shallow seismic study, energy is produced using explosives or using Vibroseis. The reflection and refraction of seismic waves help us to understand the extent and thickness of shale formation in the basin as the travel time of these waves will be different in different rocks. While carrying out shallow seismic work, it is important to observe the residential area and in the case of remote areas, animals in the forest. As the seismic activity may induce further seismicity, houses constructed with no safety norms will be the first to be damaged. In the case of well drilling, there will be large-scale pollution from the diesel and other organic emission that will directly disturb the environment of the region. In the case of a populated area or animals in the forest, it will directly affect their health. Monitoring the operation and regulating the safety norms is the most important aspect of shale gas exploration.

Construction of a drill pad and storage of water for hydraulic fracturing and flowback water cause soil erosion and contamination of shallow water aquifers. Soil erosion can be avoided by the construction of concrete waterproof cement ponds, so there is no leakage and ground water can be protected. To avoid soil erosion, a number of water reservoirs are required to store fresh water, flowback water, and treated water. Withdrawals of a large quantity of water will deplete groundwater aquifers and will impact farmers in terms of irrigation and drinking water supply.

If the site is near the forest area or inside the forest area, construction of a reservoir, drill pad, and other storage facility may result in land clearing and removal of a few trees, which will have a direct impact on our environment. The large amount of noise from the heavy equipment will disturb the animals in the forest. Emission of organic matter from the generator and other equipment will also impact the environment.

8.3 GROUND WATER CONTAMINATION

Hydraulic fracturing needs a large amount of water along with some chemicals. The volume of injected water is returned as flowback water that contains sand and silt, clay in suspension, oil and grease, and dissolved inorganic components as total dissolved solids (TDS). During hydraulic fracturing the fracking water will dissolve many salts present in the shale, and flowback water will have a higher content of barium, strontium, sodium, calcium, magnesium, and chloride. In the case of presence of radioactive material in the shale formation, that material will also be dissolved in the fracking water and come back in the flowback water.

Ground water contamination is another important aspect while carrying out hydraulic fracturing. The impact to ground water is related with shale gas exploration. Induced seismicity due to hydraulic fracturing and contamination of shallow water aquifers need to be monitored. The induced seismicity in the basin using Vibroseis or during hydraulic fracturing at high pressure can induce the fractured zone near the shallow water aquifers, and during hydraulic fracturing the fluid may migrate to the shallow water aquifers. The hydraulic fracturing and related activities can contaminate shallow water aquifers due to additives in the fracturing fluids. Flowback and produced water after fracturing may contaminate surface water. Fracturing fluid is a mixture of water, proppant, and chemical. The chemicals are added to reduce friction of the injected fluid, and the biocides are added to prevent bacterial growth. The corrosion resistance material is added to avoid corrosion.

It has been observed that chemicals in the additives migrate through the shale fractured zone to shallow water aquifers. Fracturing fluid may move up through the intersection of induced fractures with natural fractures or the space between the casings. There could be leakage and spillage of water during transport. Leakage is also possible from flowback water and wastes. To avoid any leakage, provision of proper storage is required. So far in the United States, there is no report of contamination of groundwater. There are no cases where chemicals from hydraulic fracturing fluid have contaminated shallow water aquifers.

After the hydraulic fracturing, frack water comes back as flowback water or produced water. Flowback water contains sand and silt particles, suspended clay particles, oil and grease, and organic compounds with high TDS from the shale formation. This water requires treatment for reuse or disposal to the nearby stream. The most important elements on flowback water are the concentration of carcinogenic elements like arsenic, uranium, and thorium as it is difficult to remove them through water treatment.

8.4 ATMOSPHERIC EMISSIONS

Production of shale gas contributes volatile organic carbon (VOC) and nitrous oxides to the atmosphere. VOC plays an important role in the formation of smog with NO_x in the presence of sunlight. The local community is concerned with the exhaust from the transportation on the road and vented or flared emissions associated with shale gas exploration and pollution of the environment. There could be other reasons for the pollution of the environment. Release of natural gas in the subsurface is another issue for the local community because in the case of any explosion, they will be affected. Normally, it is related with the development of high pressure due to some technical problem. Sometimes blowout occurs due to poor quality casing or cementing material. Shale gas wells have very high risk for accidents that could be related to fracking fluid or accumulation of methane gas. Monitoring and following strict guidelines are the only answer to such risks.

The emissions of gas from the shale well will have an impact on global warming. Methane has biogenic as well as thermogenic sources. Emissions of methane from the well pad may take place from a storage reservoir or even while collecting flowback water. This can be captured for resale also. A small quantity of natural gas may leak from venting valves associated with separators and condensate tanks or be produced. Some emission components including methane leakage are possible from gas processing. For economic reasons, normally, gas companies try to minimize the natural gas emissions, but a small quantity of gas leakage may take place.

Other than methane, there is emission of other gases during hydraulic fracturing from various equipment. The onsite leakage and associated air may lead to explosion. Sometimes, there is release of methane in the area, and owing to high concentrations, an explosion may occur. Methane escaping from shale gas wells may migrate in geological formations. Explosions due to collection of methane in the well can be avoided by providing venting system to the well.

Sometimes, hydrogen sulfide gas is associated with natural gas, which is one the most dangerous toxic gases. Emission of greenhouse gas in a larger area can be reduced by flaring. VOC compound is another type of pollutant that needs to be monitored for less impact to the environment. One can control the air emission or reduced the emissions using exhaust mufflers.

8.5 SHALE GAS EXPLOITATION AND HEALTH HAZARD

A large number of chemicals are added in the fracking water for hydraulic fracturing, which can be responsible for the potential health hazards. As the added chemicals are in a very small amount in fracking water the chances of health-related issues are minimal. So far, such cases have not been reported from shale gas play areas. Chemicals associated with flowback water or atmospheric emissions could have a significant impact on human health. For example, VOC impacts blood, but there is no regulation for the level of exposure or duration of exposure though even a small exposure for long time will have an impact on health.

The various operations during shale gas exploration and exploitation will emerge as potential health effects to the community living in the play area. A major organic compound harmful to humans and other living beings is benzene, as well as other volatile organic compounds (VOCs). VOCs from shale gas are responsible for diseases like leukemia, various forms of cancer, headaches, diarrhea, bleeding of the nose, dizziness, and blackouts. Some of the chemicals that are health hazards are benzene and VOCs, aromatic hydrocarbons, and air pollutants. As such studies were not carried out before the shale boom, it is difficult to assign the responsibility to the industry entirely. In fact, there is more of a threat from emission from smoking. Hydraulic fracturing has been used for the last 50 years by hydrocarbon industries, but there are no reports on persons living nearby. The major threat and contribution from exploration to environment is low, and heavy industries create more greenhouse gas, which has a direct impact on global climate, compared to carbon dioxide released during shale gas operations.

The most frequently reported observations are sinus problems, eye, throat, nasal, and skin irritation, allergies, weakness and fatigue, joint and muscle pains, breathing difficulties, sleep disturbances, swollen and painful joints, frequent nausea, bronchitis, and difficulty in concentrating on work. The presence of carcinogenic elements such as arsenic, uranium, and thorium in flowback water needs very strict regulation. Even the hydraulically fractured fluids are harmful after the fracturing, and there are toxic elements in the various salts from shale formation.

As the gas exploration/exploitation is a newer and constantly improving technique, a lot of work is still required in the play area. There is a strong need for research work related to environment and water issues. With sparse population in the United States, many problems related to health and water may not be noticed that could be very important for the densely populated countries like China and India. An increasing trend of methane gas in the shallow water aquifers in the shale play area has been observed. But specific study due to shale gas is necessary, as an increasing number of vehicles, air conditioners, and a more comfortable life are also responsible for these emissions. It is possible that fracking fluid can migrate through both existing and newly created fracture pathways. Monitoring and modeling will help to understand the nature of the fracturing process in the shale gas play area.

8.6 IMPACT ON WATER AND ENVIRONMENT

Water management is the major part of shale gas play as water is the main source for hydraulic fracturing. It is used in gas exploration and exploitation. It has been observed that the amount of water used for shale gas exploitation is equivalent to two or three days water supply used by the community in the shale gas play area. Though 100 million gallon sounds like a very large amount of water, the daily supply to the city could be 600–700 million gallons. One should prefer the recycling of sewage and flowback water to reduce the use of fresh water. This also allows reducing the disposal of treated flowback water. Flowback water after the necessary treatment can be disposed into the stream or into the injection wells. The TDS of flowback water could be very high (10,000–120,000 ppm) depending on the type of shale and its composition.

Disposal is a very important aspect under waste water management. It is a risk to the environment. Many wastes are unique to shale gas. Guidelines are required for the storage of different liquid waste. For the drilling fluids, different flowback and treated water storage tanks are required. There is more emphasis on disposal of used water through recycling and reuse of hydraulic fracturing fluids. Concentrated chemicals may be delivered just before fracturing. Necessary precaution may be taken for transport for reuse or disposal. Another potential source of water leakage is from the lining of the storage reservoir. Sometimes, even a lined reservoir can leak if the plastic liner is not of good quality. Another method is to use a reinforced cement reservoir. Waste can be also stored in closed steel tanks. Exploration has engendered a growing concern about contamination of shallow aquifers. Methane gas is a common contaminant in drinking water wells. Gas associated with shale may be of a biogenic or thermogenic origin.

8.7 NOISE POLLUTION

Noise pollution is another issue related to human beings and the environment. It may have impacts on general health. Compressors, generators, and pipe handling are the sources of noise. Ambient noise is different during day or night and in rural and urban area. For transient or temporary operations the sound level is provided under the guidelines. The various noise levels for separate operations are different. For well drilling, it will be 83 dBA, for produced water injection facilities, it will be 70 dBA, and for gas compressor facilities, it will be 90 dBA. The noise level can be reduced using a sound barrier. Under the guidelines, noise suppression on drilling rigs could be 60 dBA during daytime and 50 dBA during nighttime.

For safety reasons, industry monitors the noise level 24 h. When higher sound levels are detected, preventative action is taken.

8.8 ENVIRONMENTAL IMPACT OF BLOWOUT

A blowout is a technical snag. Following proper guidelines and using special techniques can avoid blowout. There can also be unexpected and unpredictable pressure changes due to nearby fracturing activity. In a blowout, the sudden release of gas is an impact to the environment and to the living beings in the operating area. Blowout has been reported in the hydrocarbon industry, and there is a specific technique to control inflammable gas. Impact of blowout on ground water has not been noticed.

8.9 GUIDELINES FOR SHALE GAS DEVELOPMENT

For the shale gas exploration and exploitation, guidelines should be made to prohibit the disposal of pollutant into water, for treatment and reuse, and later release to the surface water system. Strict strategies are required for the operation. Guidelines should cover the aquatic and terrestrial animals as well as migratory birds threatened from such pollution. There should be a provision to formulate guidelines about the time for release and amount of hazardous chemicals and time duration for cleaning up the spills. Most wastes can be exempted based on concentration of hazardous material. Some of the waste, though hazardous, can be exempted under the guidelines considering the economic importance of shale gas as a potential part of an energy program. There should be guidelines for clean water. Guidelines for shale development need very strict laws for various stages of operation.

There should be guidelines for using the road, as it will make the road crowded, and also because more wear of the road, which requires much higher cost of maintenance. The operator should be responsible to maintain the road in a good condition for the time it has been used by them. There should be guidelines for the drilling of the well and casing the well. Casing is required to prevent blowout. A well log should be maintained for periodic checking. Before hydraulic fracturing, surface casing should be completed and tested.

8.10 SITE DISPOSAL AFTER THE COMPLETION OF SHALE GAS EXPLOITATION

There are requirements for plugging an abandoned well that are also applicable to conventional oil. After the completion of shale gas exploitation, the site needs to be cleaned up and revegetated. Flowback water needs treatment and surface disposal. There should be regulation for the storage reservoir to avoid soil contamination. Most of the guidelines are required for the regulation of wastes generation and proper disposal. The runoff water will contaminate the surface water body, which is a part of underground waters.

8.11 REGULATIONS FOR SHALE GAS EXPLORATION AND EXPLOITATION

It is important to have very strict guidelines with a monitoring system for shale gas exploration and exploitation. It is equally important to enforce the guidelines, and a regular logbook should be maintained by the operator. A major potential for contamination is shallow water aquifers from the harmful chemicals added to the fracking water. Water balance is necessary for the optimum use by the industry and the community in the region. This will allow the smooth operation of the shale gas play, and at the same time the community does not have either any health issues nor shortage of water.

To avoid any contamination to shallow water aquifers, a Canadian company is working on waterless dry fracturing. Here, water is replaced by propane gel to create fractures in shale formation to release trapped natural gas from the formations. In this technique, gel reverts in to vapors due to heat and pressure and fracture the shale formation to release the trapped natural gas. But this technique is still in the development stage.

Under recent regulatory revisions, a focus on the list of chemicals used for hydraulic fracturing has become compulsory. Regulations are required for the effective implementation to protect human health and the environment. Monitoring of guidelines is necessary by a third party, and there should be very strict penalties to avoid any damage to our environment.

Ecological damages caused by habitat fragmentation have to be avoided. Emissions of VOC and hydrogen peroxide need to be prevented. Shale gas exploration has resulted in intensive industrial activities in rural and suburban areas, and that has resulted in social impacts ranging from noise from drilling rigs and heavy equipment to increased transportation. The wastes should be stored in closed steel tanks.

CHAPTER 9

Economics of the Hydrocarbon Industry With a Volatile Market

A.M. Dayal

CSIR-NGRI, Hyderabad, India

CONTENTS

9.1 INTRODUCTION

With increasing demand of conventional oil and gas, the prices of oil have continued to affect the growth of developing/developed countries. The oil import bills for countries like the European Union, the United States, China, Japan, and India, which are the major oil and gas users, were going up. It was necessary to develop an alternate source of energy but at par with conventional hydrocarbon or coal. Nuclear, wind, and solar energy are an answer, but they cannot replace the conventional oil or coal as the major source of energy. With increasing demand and short supply of conventional oil, the United States was the first country to finance the industry and academia for the development of an alternate source of energy. Geologists knew about the trapped gas in the shale formations, but in the absence of technology, there was no exploitation of trapped gas.

Shale gas is natural gas trapped in shale formations that are a rich source of natural gas. In the shale gas, methane is 90%, while other gases constitute the remaining 10%. Shales are a fine-grained rock and can be fractured easily. In the year 2000, the United States realized the problem with increasing prices of oil and gas and

145

Shale Gas. http://dx.doi.org/10.1016/B978-0-12-809573-7.00009-3

set up the Energy Institute at Texas to develop shale gas. With the scientific and industrial partnership, they developed horizontal drilling and hydraulic fracturing in shale formations. Principally the shale is a basic source rock for conventional oil and gas, but due to low porosity and high permeability, shale cannot store the generated hydrocarbon, and it migrates to the reservoir rocks like sandstone and limestone as the porosity of these rocks is much higher than shale rock.

9.2 OIL ECONOMY

The presence of natural gas in shale has been known about for a long time, but technology has developed in the last decade. The basic reason for the development of shale gas in the United States is that US laws grant the subsoil property rights to the owner of the land. This compensated the high cost of exploration as land was available at low rates. Large-scale pipe lines for the transport of shale gas was the another reason for its development. High initial production encouraged operators to more exploration and exploitation. Initially gas prices were $10–13 per million BTU, but they came down in the next 5 years to $3 per million BTU.

With the development of horizontal drilling and hydraulic fracturing, shale rock that was considered cap rock for conventional hydrocarbon reservoirs was acting as a source rock and also reservoir rock for natural gas. The United States was importing conventional gas from Canada, and the rates were $9–10 per million BTU, and oil was imported from the Gulf region at $110 per barrel. With the development of shale gas in last 5 years the production of shale gas in the United States increased, and prices of gas started falling; in 2014 they went down to $2.54 per million BTU (Fig. 9.1).

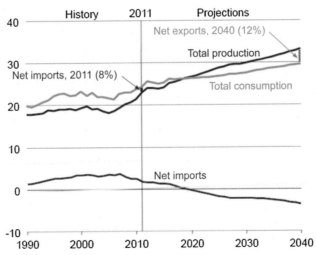

FIGURE 9.1

Total production, consumption, and import of gas in the United States (EIA, 2013b).

Table 9.1 Reserve of Shale Gas by Country (EIA Estimates, 2013b)

Countries	Shale Gas Reserve (Tcf)
China	1115
Argentina	802
Algeria	707
USA	665
Canada	573
Mexico	545
Australia	437
South Africa	390
Russia	285
Brazil	245
Total world	7299

The largest recoverable deposit of shale gases are in China, Argentina, Algeria, the United States, Canada, Mexico, and Australia (Table 9.1). The United Kingdom, France, and Poland are assessing their shale reserves. France and Poland have good recoverable reserves of shale gas, but being densely populated areas, there are environmental problems in the local community. China is taking the shale gas exploration work to a massive scale. At present, 70% of their power production is through coal, and pollution has become a major challenge for them. As they have the largest recoverable deposit of shale gas, their exploitation will bring down the environmental problem.

Though the United States has shale gas in excess, there was no facility to export it to European and Asian countries because they did not have an LNG or CNG facility for transportation at any port. In 2014, the United States government gave permission to create a facility at six ports for LNG and CNG, so it can be exported to European and Asian countries. But in the United States, there is strong opposition from the local community for exporting the shale gas to other countries as the prices of gas in the United States will increase. The United States can sell the gas to European and Asian countries at $5–6 per million BTU, which is much higher than the domestic price of $2.54 per million BTU. There is an agreement between the United States and India for the purchase of shale gas, and the first shipment has been received by Gas Authority India Limited. Other buyers in the list are countries from South America, Southeast Asia, and Europe.

9.3 GREENHOUSE GAS AND SHALE GAS ECONOMY

Overall exploration and exploitation will grow the economy of many countries. Generating the power using shale gas–fired powerhouses will produce 50% less carbon dioxide and no sulfur in the atmosphere, as it is environment friendly. Globally the major conventional sources of energy (85%) are coal and hydrocarbon.

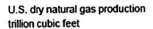

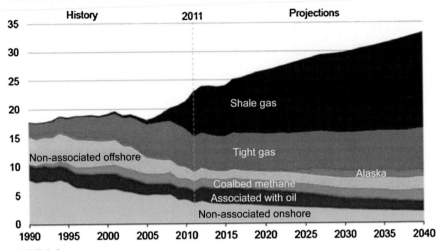

FIGURE 9.2

Projection of various energy resources in the United States. *After US Energy Information Administration, July 3, 2013a.*

Other sources of energy are nonconventional energy sources like shale gas, solar, wind, nuclear, and geothermal. Presently, the United States, Canada, Argentina, and China are producing shale gas on a commercial scale. Other countries developing the shale gas as a major source of energy are Algeria, Brazil, the United Kingdom, Australia, Poland, and Bulgaria. Shale gas and shale oil production at a commercial scale from a large number of countries will change the world's economy resulting in lower oil prices and a higher global GDP and a much better environment. Lower global oil prices and increased shale gas production will have a major impact on the future world economy. Trends of production of dry natural gas in the United States can be seen in Fig. 9.2 where shale gas and tight gas play a major role. According to EIA (2013a) sources, the United States plans to increase its production of shale gas from 23 to 33 Tcf by 2040 (Fig. 9.2). This will revive the US economy, and there will be excess oil and gas in the market. OPEC countries will not have a command on oil prices. The United States, Argentina, and Mexico will be in a position to export the shale gas and oil.

9.4 GLOBAL SHALE GAS DEVELOPMENT

With the development of shale gas on a commercial scale, many countries felt the need to develop shale gas in their countries. Prospective countries include China, Algeria, Mexico, Australia, the United Kingdom, Poland, South Africa, Brazil, and Russia. After the great success of shale gas/oil in the United States, all these countries are interested to develop shale gas as alternate source of energy in their own country. Table 9.1 shows the major 10 countries with their technically recoverable reserves. China tops the list, with a few countries in South America, North America, Australia, and Russia also ranking.

In China, Sinopec and Petro-China are the two industries working on the development of shale gas. These petroleum industries are working with the experts from the United States to produce shale gas. China has the target to develop shale gas in the next decade to replace major coal-fired powerhouses with gas-based ones to reduce the atmospheric emission of carbon dioxide and sulfur dioxide. The combined production of Petro-China and Sinopec in 2015 was 5.1 billion cubic meters. China has the largest reserves of shale gas, but due to rough geological terrain, it is difficult to provide basic infrastructure facilities like movement of heavy drilling machines, storage of water for hydrofracturing, and management of flowback water. For the economic growth of any country, it is necessary to develop an indigenous source of energy.

South America has good reserves of shale gas (Table 9.2; Fig. 9.3). The largest reserves of shale gas in South America are in Argentina and Mexico. Other countries are Brazil, Chile, Paraguay, and Bolivia. Argentina is supplying shale gas to Chile, Brazil, and Uruguay. Argentina is second after China for a technically recoverable reserve of shale gas of 804 Tcf. There are four shale gas basins that have been recognized in Argentina, and in these four shale gas basins, there are 10 shale formations that contain 802 Tcf of technically recoverable shale gas. Argentina has the advantage over the environmental issue as the shale gas reserves are in the least populated region of Patagonia and Neuquén. At present, exploration work is going in the Neuquén basin. In this basin, the Los Molles and Vaca Muerta shale formations have been selected for shale gas exploration. Vaca Muerta has technically recoverable gas and an oil reserve of 308 Tcf. Argentina produced 1.25 Tcf of natural gas in 2014. Mexico has the second largest recoverable reserve of shale gas in South America. But Mexico is not able to produce shale as there are already pipe lines for the shale gas from

Table 9.2 Reserve of Shale Gas in South America (Energy Policy Group)

Countries	Shale Reserve (Tcf)	Recoverable Shale Gas Reserve (Tcf)
Argentina	2732	804
Mexico	2336	681
Brazil	906	226
Paraguay	249	62
Chile	287	64
Bolivia	192	48
Uruguay	83	21
Colombia	78	19
Venezuela	42	11

FIGURE 9.3
Shale gas basin in South America.

the United States to Mexico. Brazil is the third largest recoverable reserve of 226 Tcf shale gas in South America. The economy and GDP of these developing countries is going to increase.

The United States has been importing oil from Gulf countries and gas from Canada since World War II. In 1970, there was a sudden rise of oil and gas prices that led to recessions, high inflation, and a slowdown in economic growth. With the increase of oil and natural gas prices in the year 2000 onward, there has been large-scale employment in the oil and natural gas industry, which continued till 2008. With increasing oil and gas prices in the year 2000 onward, the US government supported the development of shale gas. Initial production of shale gas was small, but with increase in shale gas production the prices of gas fell from $8–10 per million BTU during 2008–14 to $2.5 per million BTU during 2014–15. After the end of the recession, between 2010 and 2012, the oil and gas industry added more than 1.5 million jobs all over the United States. The exploration and exploitation of shale gas in the United States is responsible for high GDP growth in recent years. Use of horizontal drilling and hydraulic fracturing allowed more profit in shale gas and the shale oil industry, and there was large-scale employment from 2011 in the petroleum industry. With the development of shale gas/oil in the United States, the US economy has improved, and the US dollar is one of the strongest currencies in the world.

Shale gas exploration and exploitation on a large commercial scale started in 2006 in the southern states of the United States. As the industry grew, there was large-scale employment in the hydrocarbon industries in these states. Wyoming has good reserves of conventional hydrocarbon, and because of its small population, it has most of the benefits with an increase in oil price. Alaska's economy also depends on the oil extraction industry. There was rapid growth of shale gas exploration and exploitation in North Dakota. With the growth of shale gas, there was impact to the states with a large amount of coal industry like West Virginia. Louisiana and Texas have 40% of US refining capacity, but exploration and exploitation of shale gas has been an impact to the economy of Louisiana and Texas.

From 1972 to 1986 there was a sudden increase in global oil prices. From 1986 onward the global oil prices started falling, and a sudden decline in oil prices was responsible for the large-scale unemployment in the oil industry in states like Texas. There was a recession in the United States during this time. But states like California, Colorado, and Pennsylvania, which produce a large amount of oil and natural gas, could survive the downfall of oil prices. There was a big impact to the countries importing oil and gas as the prices of oil crossed $100 per barrel.

The exploration of shale gas as an alternate source of energy benefited the United States by reducing dependence on imported energy and diversifying the economy. The changes in the United States from conventional oil to non-conventional natural gas production affected the economy of the individual states. Energy Information Administration's (2015) estimates in 2013 the total production of shale gas was about 11.34 Tcf, from shale and tight oil resources in the United States, which contributes about 47% of the total US dry natural gas production.

Considering the volatile hydrocarbon market and implications of shale resources in very large amounts, it is necessary to define the reserves of shale gas and quantities that can be recovered commercially. Commercially recoverable resources are the volumes of oil or natural gas that can be produced with existing technology and can provide you profit. China has one of the largest technically recoverable reserves of shale gas, but due to their geomorphic location, all the reserves are not economically recoverable. The economics of gas is based on geomorphic location of shale formation, costs of drilling, completion of wells, the production of natural gas from an average well over its lifetime, and the global oil prices.

The economy of shale gas is influenced by global oil prices, geology of shale formations and their locations. In the United States and Canada, the land owner has subsurface rights, which provides a strong incentive to the operator. In other countries, similar subsurface rights are not available, and the land owner does not allow any such activity on their land. In the United States and China the infrastructure facilities like pipe lines are provided by the government, while in other countries, the operator is responsible for all the post-exploration activity.

In the United States, major shale gas plays are within a depth of 2000 m, while in China, they are 3000–4000 m deep, which requires greater logistic support. In first phase with the initial assessment of 11 basins in South America, 27 shale formations were selected for evaluation. The geological aspects of shale gas potential is the primary parameter. To evaluate the areas of interest in each basin the core study has to be carried out, and data is compared with shale from North America, which allows making a comprehensive geological and geophysical review of the economics of the gas play.

The amount of gas available in the basin is of prime importance. One of the most useful techniques is geochemical exploration. The core analyses for their total organic carbon content, type of kerogen, hydrogen index, and thermal maturity are the basic parameters. The formation success and recovery factor are the next parameters in shale gas play. For economic viability, it is necessary to know the technical recoverable natural gas in the play area within the prospective zone.

Shale gas exploration and exploitation can be divided into two phases. In the initial phase, it is necessary to look at the prospective area, geological condition, and access for the heavy vehicles for the drill pad development, availability of water for hydrofracturing and option to discard the flowback water. In a later phase, geochemical, mineralogical, and petrophysical studies are carried out to study the recovery factor based on geologic complexity, pore size, formation pressure, and the clay content.

Technically recoverable gas is the amount of gas that can be produced commercially with a sufficient profit margin. Based on US shale gas production the recovery factor varies from 15 to 30%. The recovery factors for shale oil are lower than for shale gas. The recovery factor is based on petrography, geochemistry, mineralogy, geology, and geomorphology. Being a new energy resource, the life of the shale wells and the exact amount of recovery cannot be determined.

9.5 IMPACT FROM SHALE GAS EXPLOITATION

In the United States with the exploration and exploitation of shale gas, the production of shale gas increased between 2006 and 2015 by 100%, and prices of natural gas came down from $8 to 2.5 per million BTU. The main industries that received the benefit of low prices of shale gas are petrochemicals, fertilizers, plastics, and resins. Shale gas production is a very fast growing industry. To produce 30 billion cubic feet of shale gas per year, roughly 700–800 wells need to be drilled yearly.

With the exploitation of shale gas the prices of gas in the international market came down 60% in the last 5 years, with the present cost being $4–6 per million BTU. Production of shale gas is responsible for a reduction in imports, resulting in a drastic improvement in the US currency. The shale gas production in the United States is partially responsible for a reduction of oil prices globally from $110 per barrel to less than $30 per barrel in 2016. Economically, there is a two-billion-barrel surplus of oil in the market, which has resulted in a downfall in oil prices. The downfall in oil prices is not only due to shale gas but also to sluggish growth of the economy all over the world. But the biggest importer of oil and gas has a major advantage with the downfall in oil prices. A country like India has a major advantage in saving their foreign exchange. One of the direct effects is a decline in import volume in recent years.

Development of shale gas definitely impacts domestic demand of coal in the United States as coal is slowly being replaced by gas as an energy source. But in developing countries like India and China, coal will be still used for major power production. The fertilizer industry is getting a benefit due to lower prices of natural gas for making urea. Another big industry to benefit is the petrochemical industry as well as refineries. Transportation, which includes light motor vehicles, buses, and trucks, can use natural gas as a fuel, which will help in reducing emissions. For the export of shale, it has to be converted into an

LNG or CNG form. In the LNG state, the volume reduces 600 times; hence, it is easy to store and transport over a long distance. From 2016 onward, the United States is exporting LNG to European and Asian markets. But the export of LNG from the United States will make the gas more expensive in the United States.

Shale gas can be used in the form of generating electrical power, which helps agriculture, transportation, and other industries. It is very good or the environment and reduces emissions as there is no sulfur dioxide in the combustion of shale gas. Development of shale gas also generates employment for laying the pipe line and other activity at the production site. It also reduces the dependency on Gulf countries for oil, and countries save a large quantity in foreign exchange. Economically, a small investment in the shale gas industry is good, though it is a part of a volatile market like oil, where the price has come down from $110 per barrel to less than $30 per barrel. Gas prices are not as volatile and remain in the range of $5–10 per million BTU. Major US companies in oil and shale gas exploration have contracted with oil companies in China and other countries for shale gas development.

9.6 CONCLUSION

Environmentally related risk with shale gas exploration and exploitation by hydraulic fracturing has yet to be assessed. This can be done only with the countries where shale gas is in a production state. So far, no concrete evidence has come for any environmental issue from the countries like the United States, Argentina, and China, which are the producing countries. Some minor issues related to water contamination and minor damage due to seismic activity have been reported. Major risk for shale gas exploration and exploitation are use of a large quantity of water and later disposal of flowback water. But there are ways to manage the water and also treatment of water before surface disposal. Earthquakes generated by seismic activity and hydraulic fracturing are minor and do not harm the community. But in developing countries, there is a strong feeling about contamination of ground water and also damage to the property due to induced seismicity. Decisions about the implementation of shale gas exploration and exploitation need plenty of information and communication with the community through media and also risk management in case of any accident.

Economically, development of shale gas exploration and exploitation will affect the conventional hydrocarbon market. With availability of natural gas at a low price and also low greenhouse emissions, the conventional hydrocarbon and coal industry are going to experience a major downfall. With the development of other nonconventional energy sources like nuclear, wind, and solar energy, use of hydrocarbon and coal will be restricted, which is good for our environment. Shale gas exploration and exploitation industries need regular monitoring from government as well as independent agencies.

References

US Energy Information Administration, July 3, 2013a.

US Energy Information Administration, May 2, 2013b.

US Energy Information Administration, September 11, 2015.

CHAPTER 10

Role of Nonconventional Shale Gas Energy in the Next Century

A.M. Dayal

CSIR-NGRI, Hyderabad, India

CONTENTS

10.1 INTRODUCTION

Fossil fuels like coal, oil, and natural gas account as major fuels for global energy consumption, while the renewables and nuclear are minor. Though the growth in nuclear and solar has been enormous, the base is small, and intermittent problems remain a significant hurdle to accept these as a primary source of energy. There are 30 countries producing nuclear power. Of these, only France, the United States, Hungary, Slovakia, and Ukraine use it as a major source of energy, although several others countries have significant nuclear power generation capacity. According to the World Nuclear Association, over 45 countries are seriously considering the development of nuclear power for energy, with Belarus, Iran, Jordan, Turkey, the United Arab Emirates, and Vietnam at the forefront. China, India, and South Korea are in the mode of expansion of their nuclear power capacities. China is aiming to increase its existing capacity to 400 GW by 2050 (from World Nuclear Association, 2013). South Korea plans to expand its nuclear capacity to 43 GW by 2030 (from World Nuclear Association, February 2013). By 2050, India plans to have 25% of its energy demand from nuclear energy. Once India joins the Nuclear Suppliers Group,

Shale Gas. http://dx.doi.org/10.1016/B978-0-12-809573-7.00010-X

it will be easy to further expand its nuclear power. France is the only country where major power generation is through a nuclear source.

Solar energy is another form of nonconventional energy replacing conventional hydrocarbon and coal as part of energy. A recent study indicates only 0.3% of the area in North Africa is generating solar energy. China is developing solar energy on a large scale as an alternate source of energy. The solar power in Japan contributes 2.5% of the nation's annual electricity demand. Thailand plans to develop 6000 MW of solar energy by 2036. India has planned to produce 100,000 MW of solar energy by the year 2021–22 under the National Solar Mission. Pakistan has set up a solar power park to produce 100–900 MW under a solar energy development program. UAE is generating 100 MW of solar energy near Abu Dhabi. By 2020, the United Kingdom has planned to increase the production of solar energy to 22,000 MW. Spain is one of the advanced countries for solar power energy in Europe and produces 6.9 TW, which is 1.7% of the country's demand. Germany's present solar energy production is 22 GW and plans to increase to 52 GW by 2025. Mexico in Latin America produces 30 MW of solar energy. The United States has an installed capacity of 27 GW of solar energy that is to be met by 2015. Australia is developing the solar power energy of 600 MW.

However, in the present energy scenario, shale gas is becoming a game changer, particularly for the United States. It is a natural gas with 90% methane and is considered a very clean fuel compared to conventional oil and gas (Fig. 10.1). As per energy information agency (EIA), there are enormous unconventional shale gas resources all over the world (Table 10.1). The growth of shale gas production in the United States over the past decade has been remarkable. With rising prices of oil and gas to improve the economy of the country, the United States took the initiative to develop shale gas exploration and exploitation. George G. Mitchell is the person responsible for the development of the shale gas industry.

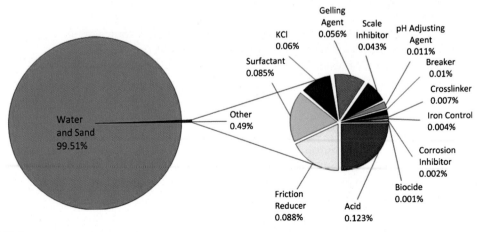

FIGURE 10.1
Volumetric composition of natural gas.

Table 10.1 Shale Gas Reserve and Recoverable Reserve (EIA, 2013).

Country	Reserve in TCF	Recoverable Reserve in TCF
China	1115	124
USA	665	318
Argentina	802	12
Algeria	707	159
Canada	573	68
Mexico	545	17
South Africa	485	–
Australia	437	43
Russia	283	–
Brazil	243	14
Indonesia	580	–

Shale formation should have high total organic carbon (TOC, 0.5–25%), should be mature petroleum source rock, and the gas should be of thermogenic origin. Shale formations should be brittle and rigid to maintain open fractures. The gas in shale is present in free and adsorbed forms and in pore spaces.

10.2 DEVELOPMENT OF SHALE GAS

A few years ago, the United States of America was importing gas from Canada and oil from another part of the world. With the development of shale gas in 2010, shale gas production was 2% of domestic output, and today it meets the entire requirement for gas. In fact, a number of coal-fired powerhouses have been switched over to natural gas to reduce the carbon emission. With large-scale production of shale gas, the cost of gas has come down from $9 to $2.54 per mm BTU. Use of gas in the powerhouses has reduced the power charges for the consumers. The production of shale gas in the United State has stopped the import of gas, and presently gas is being exported to India, Mexico, and European countries. From 2014 onward with the downfall of oil prices, the oil companies working on shale gas have been concentrating on oil production, and at present, US oil production is two billion barrels more than the UAE. With the development of shale gas/oil and tight gas, at present the United States has a surplus of oil and gas. A recent report by the National Petroleum Council projects that with such a large amount of recoverable shale gas, the United States has sufficient reserve for next 100 years.

The tight oil production in the United States was around 1.5 million barrels per day in 2011, which has increased to 2.5–3 million barrels per day in 2015. With increased offshore production, shale oil, natural gas, conventional onshore production, and the production from Arctic wells could exceed 10 million barrels per day, which will be higher than the current output of Russia and Saudi Arabia.

Shale gas has very good potential as the main source of energy all around the world. Shale gas is available in a large quantity; it is cheap and a much cleaner fuel than conventional fossil fuels. But to occupy the place of the main source of energy, the shale gas industry has to overcome the problems related with water contamination, induced seismicity, and environmental pollution. Being cheap and carbon friendly, shale gas has the potential to displace fossil fuels and emerge as an alternate renewable source of energy.

10.3 SHALE GAS RESERVES

All energy resources are related to safety and environmental risks. There are issues for the exploration and exploitation of shale gas related to water contamination and environmental protection versus economic growth of the country. There are also controversies about hydraulic fracturing, land use, methane, greenhouse gas emissions, and induced seismicity. But with the development of technology, well integrity, and discussion with the community, it can be developed to be used as an alternate and major source of energy. In France, there is a strong opposition from environment groups. But countries like Argentina, Australia, Canada, Poland, and China are ready to take control of environmental emission and increase the shale gas production to become self-sufficient for their energy basket. Development of shale gas creates skilled labor and new jobs at a large scale. In addition to the economic benefits, developing new natural gas supplies may provide a means to help countries to reduce greenhouse emissions. For a powerhouse working with a conventional fuel like coal or oil, the use of natural gas will help to reduce the emission of carbon dioxide, sulfur dioxide, and carbon monoxide. Similarly, natural gas will be very useful with low greenhouse emission in Russia, Japan, Europe, the United States, and Canada for the essential heating of houses during winters. There should be encouragement to use natural gas in a large number of motor vehicles to control pollution levels. Norway has decided to stop the use of petrol and diesel by 2020, and only gas-operated or electric vehicles will be used for transportation.

10.4 ECONOMICS OF SHALE GAS

Prices of natural gas may not always remain the same and may go up any time, which will have an impact on shale gas development. With the present rates of oil in the global market, the development of shale gas is not a profitable business. But in next few years, oil prices will be $60–70 per barrel, so exploration and exploitation of shales gas will become economical. In the future, the development of shale gas will also depend on public opinion and reduction of environmental and seismic risk. Development of shale gas is a game changer for US economic growth. The export of gas to Asian and European countries will make the US dollar much stronger in the international market. Natural gas is in many respects a clear and efficient burning fuel that has the potential to reduce carbon emissions. But due to hydraulic fracturing, induced seismicity, and emission of VOC, there is strong opposition to the development of shale gas. There is a large demand of shale gas from developing countries like India, China, European countries, Japan, and Latin America.

10.5 IMPACT OF DEVELOPING SHALE GAS

The exploration and exploitation of shale gas may contaminate ground water through the input of chemical additives in flowback water. There could be contamination of shallow water aquifers by seepage of hydraulic fracturing fluid. The emission of greenhouse gases to the environment will increase the atmospheric pollution. However, the emission of greenhouse gases from the shale gas industry is for a much shorter time compared to other energy sources. Though only 0.5% of chemical is added to the fracturing fluid, for one million liters of fracturing fluid, the quantity of chemical will be a few thousands of liters, which exceeds the safety limits as these chemicals are carcinogenic.

But as the advancement in shale gas exploitation proceeds, newer chemicals will be used that may not contaminate the well water or surface water. Normally 50% of the water comes back as flowback water, but the remaining water is within the shale formation, trapped in the subsurface. The flowback water contains the added chemicals and also toxic elements associated with the shale formation. So far, studies carried out by various US universities suggest very negligible effects to the surface water, atmospheric emissions, and community health. Presently, the contamination of ground water due to fracturing fluid has not been reported as a health issue in the United States. In next few years the issues associated with hydraulic fracturing and water contamination will have feasible solutions, and shale gas could be produced as a major source of energy from most countries in Asia, Europe, Canada, Latin America, and Russia.

The study carried out in Europe on the effect of coal-fired and natural gas powerhouses on human health suggest there are more deaths using coal as fuel compared to the using natural gas. As coal-fired powerhouses will be replaced with natural gas, it will reduce the greenhouse emissions to a large extent. Mining of coal is another health hazard, which will reduce as the demand of coal decreases. Powerhouses operated using natural gas produce much cheaper electricity with 50–70% less greenhouse emission than coal-fired powerhouses. China is the largest user of coal in the world for power production and expected to reduce coal-fired powerhouses by using natural gas as a major source of energy in the next 5–7 years. China has plans to construct a new powerhouse based on natural gas to reduce the greenhouse gas emission.

There are six liquefied natural gas (LNG) terminals in the United States to export natural gas to European and Asian countries. With increasing demand for shale gas, EIA expects a rise to come in the prices of natural gas in the United States. This should allow more exploration and production of shale gas in the United States and will also help in growth of the US economy and in the creation of a large number of new jobs.

Fracking involves drilling a well bore into the reservoir rock formation and then pumping water, sand, and chemicals into the well at high pressure to create fractures or fissures in the rock. Once the fracture is open, the released gas flows out of the fractures and into the well bore. With the increasing demand of water, alternate sources are being developed, which will solve the water-related issue. The induced seismicity is a short time event as it is related with drilling and hydraulic fracturing. Many labs are working to replace water as the fracking material with another material like ethane gas. In the future, additives will be developed that will not be harmful to the land and aquatic life. Well blowouts will be reduced with monitoring of subsurface gas pressure and a required safety vent will be provided. Disposal of waste water will not be an issue in the future as water is not going to be used for fracking.

10.6 SHALE GAS RESERVE AND PRODUCTION

For electrical power generation the main sources are coal, oil, and gas, which are less expensive and more abundant, but they are also the main sources of greenhouse emissions. Renewable energy, such as solar, wind, geothermal, and biomass power generation, is more useful as there is no environmental effect, but it cannot replace coal or conventional hydrocarbon. Recent development of shale gas in the United States indicated that shale gas is the alternate source of energy for coal and conventional hydrocarbon. In 2008, the United States was importing a natural gas supply, and in 2016, it is ready to export LNG to European and Asian countries. The United States is going to become a significant global player in the natural gas market.

Canada is the world's third largest producer of natural gas, with an average annual production of 6.4 TCF. At present, Canada's production is less than the United States for natural gas, but with the decline of conventional natural gas sources, industries are turning to unconventional shale gas. Many companies are now exploring and developing shale gas resources in Alberta, British Columbia, Quebec, and New Brunswick. Initial drilling at Quebec's Utica shale reserve suggests that the reserve could hold more than 20 TRC of recoverable gas. Canada currently exports about 50% of the natural gas it produces, with no demand from the United States. Canada is developing other markets, and the industries are investing in the necessary infrastructure in northeastern British Columbia. This terminal would allow Canada to export LNG to Japan, South Korea, and China.

Preliminary exploration in South America suggests that good shale gas reserves are present in Argentina, Brazil, and Colombia. Argentina is the only South American country producing shale gas in the Neuquén Basin. Most of the shale gas projects in Argentina are being undertaken as joint ventures. The development of shale gas in Argentina will be valuable to the country.

Large reserves of shale gas have been reported from the United Kingdom, Netherlands, Poland, Germany, France, Scandinavia, and Norway. Exploration activity is primarily through joint ventures to share risk and know-how. Compared to the United States and Australia, except for a few countries like the United Kingdom and Poland, other countries are not very keen to exploit their shale gas reserves in view of water contamination and environmental issues.

In the United Kingdom, shale gas production has commenced at the Blackpool aquifer in Lancashire, and new shale gas reserves have been found in the Mendips. Production at the Blackpool aquifer was voluntarily suspended due to induced seismic activity. In May 2011, contamination of shallow water aquifers was discussed in UK parliamentary committee. The committee concluded that, based on estimates of shale gas reserves, exploration and exploitation of shale gas is necessary to make the country more self-reliant and to reduce the import of natural gas.

In Eastern Europe, Poland has a good reserve for shale gas and is active to exploit it, while Turkey and the Ukraine have some potential. Russia has very good reserves for conventional gas, and a major export market is China. Russia's national oil and gas company has signed a strategic partnership with Exxon to transfer shale gas–related technology, which indicates that Russia has planned to exploit shale gas reserves in the future.

10.7 CONCLUSION

Shale gas is emerging as the fastest growing source of power in the global energy basket. It is the fastest and cleanest fuel. Nonconventional shale gas or oil will have a much better market in the next decade. The United States shale boom diminished OPEC's long lasting dominance on the global oil markets. Now with excess oil and gas in the global market, prices of hydrocarbon will not escalate beyond $70 per barrel. As of the middle of 2016, oil prices were in the range of $45–50 a barrel. In next few years, oil prices will be in the range of $50–60 per barrel. This will allow the shale gas industry to grow in the next few decades. The United States shale industry has proved that it can remain in the game in spite of low oil prices. A report published earlier by Bloomberg Intelligence claimed that a lot of shale patches in Texas were profitable, even at crude below $30 per barrel. In fact, the report stated that Eagle Ford formation in DeWitt Country had an average profitability at a crude price of $22.52 a barrel.

In the next decade, renewables and nonconventional energy sources will have a larger energy demand. The United States has decided to reduce the use of coal by 220 million tons of oil equivalent by replacing coal with shale gas. But some of the developing countries like India and China will still depend on coal as a major source of energy. Looking at the global market, IEA predicts a global rise in gas production over the next two decades, and nonconventional gas will play an important role.

China, with a reserve 1115 TCF of shale gas, has the potential to be self-reliant for nonconventional shale gas. In a global scenario of economic growth and huge energy demand in the next decade and also in consideration of the environmental issues, shale gas is going to replace the main source of energy. Demand of shale gas will increase as the technology for the shale gas exploration will improve. With the large reserve of 6609 TCF of shale gas globally and the demand of 106 TCF per annum, the existing reserve will provide energy for the next 60 years. In fact, Norway has decided to allow only gas-based or electric-operated vehicles by 2025. The demand for compressed natural gas or electric vehicles is going to increase in the future. There is a good subsidy for electrical vehicles to reduce atmospheric pollution.

By 2035, major shale gas production will take place outside the United States. The shale gas in the United States will keep growing in next decade, and the production will reach 80 billion cubic feet per day over the next 20 years. The United States is the largest producer of shale gas, which will serve for the next 40 years. China has the second largest reserves of shale gas, and with current demand, it will be good for the next 400 years. India's present reserves are good enough for 50 years, which does not include all the shale formations. With increasing emission of carbon dioxide and sulfur, shale gas is going to replace, mainly, the coal-powered thermal powerhouses in the next century. Though there is atmospheric emission of methane, contamination of shallow water aquifers, and essential requisite of management for flowback water, these are the problems that can be solved. Vast shale reserves all over the world will replace conventional hydrocarbon and coal in the next decade.

References

Schneider, M., Froggatt, A., Hosokawa, K., Thomas, S., Yamaguchi, Y., Hazemann, J., July 2013. The World Nuclear Industry Status Report. Paris, London.

US Energy Information Administration, May 2, 2013a.

Index